'Unlike anything I've ever read. A tender and bewildered account of motherhood, and a powerful reflection on what it means to be alive in the Anthropocene. There is so much beauty here – and so much uncertainty – as well as an extraordinary capacity to inhabit both conditions at once. Jukes' voice is cracked open to the world, but to be cracked open is also to be joined to others, not only in exposure and risk, but through care and solidarity. The achievement of this book is to widen those circles of care – and to situate human experiences of mothering within the vast unfolding story of our animal neighbours and kin.'
Michael Malay, author of *Late Light*

'*Mother Animal* wrests motherhood from the clutches of the patriarchy and gently places it back into the hands of birthing bodies, human and non-human. A dream blend of exquisite personal memoir and compelling environmental writing, I gobbled it in one sitting.'
Sally Huband, author of *Sea Bean*

'Jukes doesn't mess around. She has a genius for getting straight to the heart of the matter. And when the matter is – as it is in *Mother Animal* – the whole business of how to live properly as a human, you simply must hear what she has to say. It will change you.'
Charles Foster, author of *Cry of the Wild*

'A deeply thoughtful interrogation of motherhood and the way it ties us to our natural (and not-so-natural) environment.'
Leah Hazard, author of *Womb*

'This beautiful exploration of pregnancy and motherhood brilliantly and powerfully explores and expands our understanding of the mother-nature / nature-mother space. By breaking down borders between human and animal "mothering", this wonderful, profoundly personal journey in and out of self leaves us feeling at once wilder and, somehow, more human.'
Rob Cowen, author of *The North Road*

'A beautifully written exploration of pregnancy, birth, and parenting. With an emotional precision, Jukes stitches human and animal lives into a vivid tapestry that is as moving as it is illuminating. I found solace in connecting to my animal self. A book to devour.'
Joanna Wolfarth, author of *Milk*

MOTHER ANIMAL

HELEN JUKES

Elliott&Thompson

First published 2025 by
Elliott and Thompson Limited
2 John Street
London WC1N 2ES
www.eandtbooks.com

ISBN (hardback): 978-1-78396-838-1
ISBN (trade paperback): 978-1-78396-902-9

Copyright © Helen Jukes 2025

The Author has asserted her rights under the Copyright, Designs and Patents Act, 1988, to be identified as Author of this Work.

All rights reserved. No part of this publication may be reproduced, stored in or introduced into a retrieval system, or transmitted, in any form, or by any means (electronic, mechanical, photocopying, recording or otherwise) without the prior written permission of the publisher. Any person who does any unauthorised act in relation to this publication may be liable to criminal prosecution and civil claims for damages.

9 8 7 6 5 4 3 2 1

A catalogue record for this book is available from the British Library.

Typesetting by Marie Doherty

Printed by CPI Group (UK) Ltd,
Croydon, CR0 4YY

CONTENTS

	Author's Note	vii
I	The Enclosure	1
II	Birth	23
III	Animal Formulas	53
IV	Forever Milk	81
V	Nests	123
VI	Communities of Care	151
	Acknowledgements	183
	Bibliography	185

AUTHOR'S NOTE

Memory is by nature friable, and during periods of extreme sleep deprivation profoundly so. This book is an attempt to mark a path through that uncertain territory while remaining true to the emotional tenor and my own recollections of that time.

For reasons of anonymity I started out by changing the names of some of those who appear in these pages, but as an approach this soon felt inconsistent, so in the end all names have been either changed or omitted.

A note on terminology: I have tried to distinguish in the text between the 'animal mothers' dreamed up by Western science, and 'mothering creatures' – a term that encompasses a whole array of parenting experiences from the natural world, some of which are still in the process of becoming known to researchers. Though I describe my own journey as a woman who carried my child through pregnancy, and while I am particularly curious about how the experience of female parents in Western, patriarchal society may be circumscribed by ideas of Nature and naturalness, I hope that the emergent truth of this book is one of diversity and change. Not all mothering creatures are female; not all are biological parents; many act as part of wider communities, in complex relationships with other individuals and species.

This being a personal and sometimes circuitous story of surprise and discovery, it does not and could not offer an exhaustive study of parenthood in the animal kingdom – that work is best done by others. New research is emerging all the time, and with luck someone else will soon write another book about animal parenting, and then another. For now, here is what I learned, and if there are grounds for alarm in what is contained here, I hope there is also much to inspire wonder.

I

THE ENCLOSURE

The month I found out I was pregnant was a record-breaking one – the planet's hottest since measurements had begun. It was July, and we were caught in a blistering heatwave; a pocket of high air pressure stretching from western Russia all the way to the Atlantic. Everywhere there were stories of desiccation and sudden deluge. In Berlin, police were using water cannons usually trained on rioters to cool the city's trees; in France, they were cautioning the public against diving in unsanctioned pools. In Greece, helicopters had been brought in to stem wildfires that had whipped through sun-baked pine forests and over roads and into houses, sending inhabitants running for the sea from which the firefighters flew. On a farm near us, thousands of chickens had just sweltered to death when barn ventilation systems failed – an uneasy foreshadowing of the nearly 900 people in the UK who would die as a result of the heat that summer, most of them not out in parks or on beaches but in their own homes, which had overheated.

Our house, a small worker's cottage built nearly four hundred years ago with thick limestone from the surrounding hills, held on to the cool – even in midsummer it was a place of shady corners, of surprise relief, the sash windows pulled down at the top and up at the bottom to catch any breeze that passed.

Outside, the earth hardened and the grass bleached. I went for walks in the early mornings and evenings, and if I passed others on the footpaths, we spoke of little but the weather – our tones initially elated and later, increasingly, alarmed. Inside, there were surprise infestations: ants teeming from the skirting boards, a wasp's nest blooming over the back door. The insects were armoured and manifold, aggressive in their work of procreation – I, on the other hand, couldn't seem to do a thing. I felt clammy and then swollen, as though I were retaining moisture somewhere, which I suppose I was – a thin layer of cells in my uterus having formed a filmy sack in which water and electrolytes and a small cluster of cells now floated and multiplied.

I thought at first it was the weather making me sick. I wasn't used to the temperatures, the long hot days. I'd missed a period. But perhaps people missed periods in the heat? Still, I drove into town for a pregnancy test, returning pinked and nauseous, pupils shrunk to pinpricks, blinking. Squatting over the toilet, I fumbled with the plastic pen, blindly waving the flimsy thing as urine mixed with the paper strip and with the damp sweat on my palms.

A blue line. The colour of water, oceans, incontrovertible facts. The radio was on in the kitchen. There was talk of thunderstorms.

I've read that some escapologists and deepwater divers slow their heart rates and manage their fear by 'remembering' being in the womb – a time when mouth, nose, ears and lungs are filled with liquid, and we possess no fear of being submerged in water, or of being without it.

That night, the night of the pregnancy test, I remember the house looked different. I was suddenly aware of its imperfections: the cracks in the plaster, the loose wires hanging from the ceiling and the damp patches along the walls. The place seemed shaky somehow – full of waiting hazards and jobs that, since moving in the previous winter, had gone unfinished.

We sat at the table, making a list of tasks to be completed over the coming months. —We'll have to fix the floorboards, I was saying —and we should repair the windows, and get the boiler serviced, we might have to replace it, don't you think?

My boyfriend glanced in the direction of the window. Outside, the sky was cooling, the blue deepening, the heat lifting slowly from the hills. —Are you listening? I said. —Sure, he said. —Sure, the boiler service. Have you noticed the ants are back?

And there they were, visible as a patch of brittle-looking movement beside the skirting board. He frowned. —I'll look it up, he said. —I'll look up what to do.

And with that, we returned to our phones – he to his pest control measures, and me to the news. *The Guardian* had done a photo feature on how animals were coping with the heat. There were pictures of zookeepers rubbing sun lotion onto the backs of tapirs and feeding ice lollies to polar bears; of farmers rescuing fish from lakes that were fast disappearing. I suppose this was intended to offer some comfort: look here, at Nature's creatures! Look at these creatures *surviving!* And at these enterprising humans, still capable of rescue despite everything. I scrolled, clicked, scrolled again, and came across a video someone had posted of a bonobo mother in a zoo not far from us, sheltering her infant from the midday sun.

I remember feeling very keen, as I watched this film, to deduce the breadth and type of her enclosure. A climbing frame, a patch of grass. Fences, walls. The mother cradled the infant as she roved from one patch of cramped and crowded shade to another. But she looked nervous, I thought. She looked visibly stressed. She couldn't escape, was the thing. She couldn't escape the heat.

What the internet suggested we do about the ants:
Poison them
Pour boiling water over them
Hoover them up
What happened when I hoovered them: more came. And by September the inside of the hoover contained a layer of furry little carcasses gently decomposing inside the plastic.

The first trimester, then: sticky, nauseous, a strong aversion to most tastes and smells, a sudden desire to disinfect everything, foggy-headedness, tense hope.

THE ENCLOSURE

The word nausea comes from the Latin *nausea*, meaning seasickness, and from the Greek *nausia*, meaning disgust and – literally – ship-sickness, but in English the word has always held associations beyond oceans. In nausea, it is possible to be both at sea and landlocked; to inhabit a body utterly persuaded that all taste, all touch, all outside stimulation is utterly, incorrigibly detestable. You long for the world to become still, for all movement to stop – knowing as you do that the source of your problem resides not with the world but your own insides, which have conspired to hold you like this: confined, desperate, unable to stop *feeling*.

Standing shakily in front of the bedroom mirror, I scoured my body for signs of change. Was my left breast not slightly fuller than last week? Was there not a new roundedness, now, to my middle? Was it OK to want this, while fearing for its future?

It seemed unthinkable that I, my body, this taut and nervy frame, might possess the practical wherewithal to gestate and birth another being. Yet if this was truly happening, it appeared to be proceeding in a surprisingly haphazard way. Discernible changes were not limited to those parts of myself where I had assumed gestation took place, but instead proliferated wildly, erupting in sudden and increasingly bizarre ways: tears at bedtime; light-headedness in the shower; new dark hairs springing from around my ankles and upper lip. *What was I becoming?* During pregnancy, the singer Adele reportedly grew a beard. 'I call it Larry,' she told a magazine, as though in coming to motherhood one might birth not just a baby but an alter ego – a second self. (Did Adele discover too that, in the months after childbirth, a mother's voice deepens by as much as a piano note? That the reverse happens outside of pregnancy and around the time of ovulation, when voice pitch increases, since the hormones behind egg release also have a hand in voice?)

I bought a foetal development chart and hung it up in the kitchen. The chart broke pregnancy down into forty pages and forty weeks; each week, a picture of a different fruit corresponded to the size of the growing foetus.

Six weeks: a pomegranate seed.

Seven weeks: a blueberry.

The delicious horror of skipping ahead – imagining oneself harbouring an aubergine, a watermelon.

By now I'd dipped into pregnancy websites and learned the dos and don'ts by heart. Do rest, eat plenty of fruit and vegetables (but be sure to wash them first), exercise (but nothing too strenuous) and *trust your instincts*. Don't eat raw meat, unpasteurised milk or cheese, uncooked eggs or shark or swordfish; don't drink alcohol; don't inhale cigarette smoke or some paint fumes; avoid dry-cleaning fluids, cat litter, hair dye and overly hot baths. Also, use your seat belt. Also, don't be anxious.

So the air I breathed contained petrochemical fumes that increased the risk of miscarriage; the soil on a carrot could contain parasites that *could* cause foetal brain or liver damage, or miscarriage. And what if on occasion I forgot the rules? What if I misinterpreted them, or misplaced them, or ate a cheese I shouldn't? *Miscarriage!*

I was not just a vessel but a membrane – a thinking, feeling boundary between my unborn child and the rest of the world, both at the mercy of whatever threats were at large in my environment and locked in an urgent, impossible struggle to control it. I began peeling mushrooms before eating them. I ordered an organic veg box, roasted a cauliflower for the first time, spent long minutes scanning the ingredients on food packets in the supermarket. Was it still OK to reheat old rice? Was it less OK than before? And all this in service to a different kind of foreignness – a body of cells, now person-shaped, steadily blossoming on my inside.

At twelve weeks the foetus was passionfruit-sized, and I had learned to eat crisps to stem the nausea, and that the development chart I'd hung in the kitchen was full of shit. Foetal size varied, of course it did – so did vegetables and fruit.

At twelve weeks, the *roughly* passionfruit-sized foetus had developed rounded buds that would eventually become teeth; the basic structure of a skeleton now hardening into bone. Occasionally, unbeknownst to me, it made a jagged little movement – spasmodic expressions of joints, limbs and growing musculature.

I lay on a bed with a plasticky feel as a bearded man in blue hospital scrubs lifted my trouser band up and lowered it, and I watched as beads of light on a screen above my head moved in and out of a human form. A bowed head, a coiled leg. A tiny, too-perfect fist. The man clicked quietly at his computer. But something was wrong – the limbs were frozen, motionless. He hadn't noticed. How had he not noticed?

—It's not moving, I said, growing panicky. The sonographer looked up, surprised. —It's a photostill, he explained gently. —Look. See that? He held a cursor over a thick dark region in the top right corner of the screen. —There.

It looked to me like a void. —What is it? —Your placenta.

I squinted. I wanted him to point again to the legs, the feet, the little alien head. I tried to recall what the placenta was, what it did. He twizzled his probe and prodded and the creature inside me on the screen gave a jerk. I had, I realised, very little understanding of what actually in a physical sense was going on.

—

When first attaching themselves to the walls of the uterus, placental cells disguise themselves as uterine cells, convincing the body that they are a part of it although in fact they are not a part of it, or not exactly. Exactly, they are something different: a fast-growing and completely novel organ with the sole and finite purpose of gestating another body, another being. The deception is necessary because without it the body would identify the cells as foreign and destroy them.

The placenta connects, then, while keeping separate, supplying the foetus with nutrients and oxygen, removing waste and producing hormones that help regulate interactions with the maternal body. Looking it up, I was diverted – found myself on a page for veterinary science, scrolling through diagrams of placental forms. *Diffuse, cotyledonary, zonary, discoid*. They were the shapes of croissants, bao buns, deep-sea worms – irregular, ragged, flaccid-looking things, more reminiscent of abstract art than science textbooks, or bodies.

Cow placentas are tufted. Dog and cat placentas form a band like a seatbelt around the amniotic sac; so do elephants'. In some form or other, the placenta has evolved multiple times, along otherwise independent branches, in every vertebrate group except birds – and so I had this curious, just-made organ in common with expectant boa constrictors and water voles; with some sharks and rays and bony fish. All species who at some point in history began retaining embryos inside their bodies and releasing their offspring whole.

Longer gestation times offer greater protection for developing offspring, but they are riskier and more resource-heavy for the mother, who also, literally, gets heavier. Bats fly slower and lower in late pregnancy; the dolphin experiences a 50 per cent increase in swim drag. The foetus of a humpback whale reaches such epic proportions inside her that it comes to function as an acoustic object – a symptom of late pregnancy, as Rebecca Giggs puts it in her book *Fathoms*, is 'a shift in the timbre of the mother's voice'.

While speaking on the phone, I often doodle on the edge of my diary, repeating little circles or squares or misshapen stars, and at a certain point in pregnancy these doodles turned to placental forms.

The human placenta, a discoid shape, averages a thickness of around two and a half centimetres, and by the end of pregnancy may weigh somewhere in the region of 500g. We tend to speak of it as belonging to the mother, but this gives a misleading impression – it is made up of embryonic cells as well as maternal ones. It is, in part, the creation of a person still in the process of being formed. Burying deep into the womb lining, it taps into bodily tissues and the wider arterial system with such forceful invasiveness that one can't abort a human pregnancy without risk of lethal haemorrhage.

My doodles were shaped like boats, and then like branches – the intricate vessel structure covering the placenta's foetal side. Sinuous and extensive, dividing again and again, I doodled it into the margins and across the days of the week as I spoke with students about their end-of-year projects; as I chatted with friends about summer plans. I'd told a few of these friends, by now, that I was pregnant. —I'm not sure what to do, I'd told them. —Should I be preparing or something? I don't know how to make it feel real. Hearing this, they'd been laid-back. —Just do what feels natural, one of them had said, and I'd come away confused, unsure if I felt natural or not.

It was around this time that I found myself returning to the film of the bonobo mother. *Protective bonobo mother takes care of newborn amid scorching heat*, the caption read, and what struck me now as I watched it again were the multiple narratives this triggered. The racketed temperatures, the scorching heat, and yet the vision front and centre of the devoted mother. What had the person filming seen here – what were they inviting me to see? A glimpse of a world out of kilter, or one of Nature's resilience, its capacity to carry on regardless? Perhaps both. Yet this put a lot of pressure on the figure of the mother, who alone appeared responsible for maintaining an image of naturalness, for 'taking care', even as the air around her boiled.

The bonobo moved, restlessly, under her block print caption. I found that I was interested in her, in what she was like. I imagined her imbued with maternal instinct, with innate knowledge about what to do. I understood that I would soon be imbued with this too, though I had little sense of the means or timing of its arrival. I'd read about waters breaking. If this didn't happen naturally (that word again), I had been told that a doctor would end up doing it for me; doing it to me. I pictured a metal implement (a fondue fork, a fishing hook) inserted rudely up my vagina; an ensuing pop, like a water-filled balloon, like the water bombs we made as kids, whereafter I would be hit – *smack!* – by the, my, instinct. (And what would I feel at this point? Release? Pain? Blind terror?) Yes, it seemed that one could either prepare or not.

—

In autumn, we made our way through the DIY jobs. I bought a pot of wood filler and began repairing the window frames, seeking out the little nicks and cracks where moisture builds, digging out the rotten wood and scraping the hollows, then filling them with a putty that dried as quickly as I worked.

We packed insulation into the dust-filled loft; patched up the plaster in the bedroom. Soon a man came to mend the tiles on the roof; another rewired the kitchen. Meanwhile I cleared out the shed and pulled at weeds in the garden, uncovering as I did the evidence of other creatures: packets of vegetable seeds bitten into by mice; woodworm in a cupboard door; slug trails winding around the plant pots.

One morning, pulling a cabbage from the veg box and stripping its mottled outer leaves, I came across a clutch of eggs – tightly packed, anaemic-looking. A larva sprung and coiled. I yelped. Reached for a large kitchen knife and lopped them off, shocked at the sight of this secret work, this mysterious and strange labour that some ancestor of mine had long ago selected to keep inside them.

At sixteen weeks the foetus was avocado-ish sized, and the extra heart inside me was capable of pumping over 28 litres of blood a day. The diaphragm contracted and relaxed, contracted and relaxed, submerged in its salty fluid. The hands flexed, formed fists.

—Did you know, I asked my boyfriend, wanting somehow to involve him in the physical experience. —Did you know, I said, being pregnant is like running a forty-week marathon? Did you know the feet get bigger during pregnancy, as does the heart?

Did you know became my mantra. As in, you'll never believe. As in, let me amaze you. The world is extraordinary. The world is extraordinarifying. If the mother doesn't get enough calcium in pregnancy, the foetus will begin drawing it from her bones.

The bump was getting bigger. The kind of bigger that prompted polite, weary-looking men to get up from their seats on busy buses; prompted people to carry things, open doors.

My friend Dan, visiting the house, spotted the development chart hanging in the kitchen and remarked that it might be funny to make one measuring foetal size in relation to electronic devices.

Seventeen weeks: a smartphone.

Twenty-two weeks: an iPad.

—But no one would, would they? Dan said. —No one would buy it.

We sat on the sofa and picked slices of mango from a plastic packet. —Anyway, it'd be back to front, he said, stretching his legs. —Smaller is better in technology terms, right? I mean, you'd begin with nanotech and end up with what, a PC?

He had a point. Who wants to imagine their unborn child in terms of lithium and silicon and polycarbonate parts? We want them untainted, I suppose – and to preserve the innocence of that floating, uterine world.

The chart was not just about size, then, but association. A healthy foetus was Nature's fruit – satisfying and wholesome!

An image that confusingly brought to mind ingestion, not gestation, so that my ritual on turning a new page each week was to go out and buy the pictured fruit.

Dan had a daughter of his own, and she'd arrived early – they didn't even have a cot or a car seat ready. —Be prepared, he warned solemnly, as he left the house that day, so after we'd said goodbye I resolved to pull out my laptop and get to it.

A maze! A vast, sprawling maze of safety features and comfort ratings and special one-day discounts. I scrolled and I scrolled, and was snagged, and thrown into blind, covetous comparisons, and backtracked, and scrolled again. There were algorithms, clearly, and a lexicon – a language through which sellers communicated with people like me. With the type of person I'd led them to believe I was. The natural world, as it happened, featured heavily. Soon I'd picked up a pen and begun making a list of keywords: *pure*, *simple*, *organic*, *natural*; *plastic-* and *cruelty-free*.

There were biodegradable baby wipes and bamboo bowls; organic swaddle blankets and wooden rattles. The Nature on offer here evidently didn't include the bacteria or parasites I'd been working so hard to protect myself from. No, this version was clean – serene, even – and pricey. Smiling mothers were pictured gazing at happy, perfectly formed infants. Here, it seemed, were the fortunate bodies: the ones that had succeeded in keeping out the bad Nature, and locked the good Nature in. But, I thought, the images were flat. Yes, there was a kind of monotony to them, perhaps because the version of naturalness (and indeed of motherhood) being put forward here felt strangely closeted.

I looked at the picture in front of me: a wooden mobile suspended over a baby's cot. A carefully counterweighted arrangement of circus animals that one could imagine might bob and wobble pleasingly if placed beside an open window. But there was no window in this picture; there was no door. Nothing to suggest that a world even existed beyond the neutral tones of the nursery.

The animals dangled. The baby gazed. The suggestion seemed to be that in motherhood my range of focus would begin and end with what was directly in front of me; that my role lay in simulating this rather bland image of simplicity and calm.

I blinked at my shopping basket: three packs of wipes and a tub of nappy cream. The postman appeared at the window and jammed a bunch of letters in through the box, then sat a moment in his van.

At the sixteen-week scan, having asked if we wanted to find out the sex, the sonographer had proceeded to spend fifteen minutes hunting around for a penis – whereupon, unable to find one, he'd concluded that I was most likely carrying a girl. —That'd be my strong guess, he'd said. —That'd be my theory. Thus was her sex assigned in the first instance on the basis of an absence, a lack.

Now I smoothed my hand across the taut skin of my stomach. *No*, I thought. Not a lack, not here, not so far as I could tell. The foetus lunged and twisted and I kept my hand there, unsettled by the nursery, which was supposed to lure me in. Too much was occluded; too much was missing. I did not want to enter that room with its inert animals, its lack of air. I didn't trust it.

Outside, the postman took a bite of sandwich, started up his van. Above, clouds bowled. The sun shone, and the bright green of the trees made me think of what secrets might be contained within the folds of cabbage leaves.

—

—Did you know, I asked my boyfriend, as I was lying in bed and he was getting dressed. —Did you know, I said —when the Surinam toad has laid her eggs, after the male has fertilised them, he lifts them onto her back, where a thin film of skin grows over the top, which the babies break through when they emerge?

—Huh, he said, pulling on his socks. —No, he said. —I didn't know that. What's a Surinam toad?

—Did you know also, I went on —some cockroaches carry their eggs around inside an ootheca, an egg case, which they attach to the side of their bodies, like a purse? And that the cichlid fish carries her eggs around inside her mouth? That the cuckoo catfish lays her eggs in the cichlid fishes' nest, so the cichlid will do the brooding for her?

He'd padded off downstairs. I could hear him moving about in the kitchen. It was windy and raining outside; the lane was a muddy stream, scattered with sticks and stones and pieces of fallen tree. The seasons were shifting, and the nausea was fading – yes, it was definitely fading, and in its place new appetites formed. I wanted stories of motherhood's deviations, its hidden transgressions and silent vastnesses; its weirdness and its wonders.

Female woodlice have a brood pouch similar to kangaroos; kangaroos have two uteruses and three vaginas but never grow a placenta. Their offspring emerge half-made, in an embryonic state – blind, deaf and bald.

Start off down a path like this and it soon becomes difficult to stop. Mice and rats have two separate uteri and a cervix for each. Tawny sharks also have two uteri and embryos capable of swimming between them – sometimes offspring poke their heads out of their mother's cervix and glimpse the world beyond even before the point of birth.

—

During the summer heatwave, a curious phenomenon: a temporary spike in deaths by drowning. In the news, these are known as 'bathing accidents' – a term that gives them an absent-minded feel, like leaving the tap running, like dropping the soap in the bathwater. Not like the more dangerous impulse to be submerged, to go under – a state that most closely resembles our time *in utero*, except that there is no return to that place; there is no going back.

For anyone finding themselves trawling the internet for stories of animal pregnancy, there's a particular meme you're likely to come across again and again: *The amniotic fluid of all mammals is remarkably similar to seawater.* It's quoted on holistic therapy pages and natural birthing sites, environmental campaign pages and nature blogs – the claim so ubiquitous as to have gathered its own momentum, its own truth. Sometimes, it comes accompanied by a coda: *Amniotic fluid mimics the seas that nourished our ancient ancestors.* How easy it would be, how neat, if we possessed this physical signature – this salty marker of our first origins. Instead, in actuality, a slight mismatch: amniotic fluid has a salinity of about 2 per cent; the oceans around 3–3.5 per cent.

A closer parallel might be this: by late pregnancy, amniotic fluid contains sugars, scraps of DNA, fats, proteins, foetal piss and shit. It holds life's first debris – its first waste.

―

—Are you feeling connected with the baby?

Running through her list of questions, the midwife squinted at the words on her screen. I drew my feet under the chair, hooked one behind the other. —What, I asked —like emotionally? The question seemed mildly ridiculous. How to experience a connection with a being that was not yet separate from me? Her finger hovered over the computer key. There was clearly a right answer, and a wrong one. I paused. She glanced up; a look of kindness, or understanding, or pity. —Talk to her, she told me. —It's good for developing the bond.

She ripped a square of paper from a large roll by the desk, laid it down on the bed next to her and gestured for me to climb up. The paper crinkled under my thighs as she took a tape measure and stretched it from belly button to pubic bone. —They hear everything, you know. Heart beating, stomach growling, air moving about in your lungs. She bent down to check the measure. —Whatever noises are going on around you. By now she'll be able to recognise your voice. They did a study that proved it. Foetal heart rate quickens when the mother speaks.

Did you know, I whispered to my unborn child. *Did you know, about the burying beetle?*

The burying beetle *Nicrophorus orbicollis* builds its nest first by finding a dead meat carcass, then dragging it underground. Next the female lays her eggs on or inside the dead body, which might be – have been – a small rodent or a bird. Once hatched, the larvae are passed morsels of carrion by the parents, who chew the meat and partially digest it before feeding it to their young.

If they'd started out as escapism, these stories of mothering creatures were quickly developing the quality of something more urgent and necessary – of matter pulled, lightning quick, below ground. They were not about research in any straightforward sense. Instead, they melded with other thoughts; they mainlined seams of feeling, situating themselves firmly as part of the process – part of the preparation.

Scientists did a study recently in which burying beetle larvae were divided into two groups. One group was fed carrion that had been chewed and regurgitated by the parents, and one was fed carrion that hadn't. The survival rate of the first group was significantly higher, the parents' saliva apparently containing some kind of protective mechanism, not unlike the function of mammalian milk. This exchange of fluids, which persists long after birth, appears to be an essential component of parental care.

———

By March, I tired easily. I moved slowly, I took longer to do small things. I was constantly hungry, but struggled to eat much; my stomach filled too quickly, and my bladder was always overfull.

We were nearing the end of our DIY list. My boyfriend had started painting, and I'd been ordered out of the house – because of paint fumes, he'd said. I called him from the supermarket. Behind me, a woman pushed a screaming toddler in a shopping trolley.

—What the hell is that? —I'm on the fruit aisle, I told him. —Do you want cantaloupe or honeydew?

He was on the stairs. He was painting them dark grey. —I'm just doing the edges, he said. —I'm nearly finished. Cantaloupe?

I thrust the fruit into my trolley as he proceeded to ask for three things from the other end of the shop, its farthest wall.

Outside, in the carpark, the sky was clear. Pigeons lumbered over the tarmac, hurried by the wind. I stood a moment, feeling the air; the first gusts of spring. Here's another thing I learned: the human brain changes in pregnancy. Key areas actually *lose* grey matter, in a process not of stupidification but streamlining. The brain spring cleans; it pares the parts it is most going to need, pruning neural connections so as to increase their efficiency. The areas selected are those related to emotion, bonding and social contact.

Back at home the stairs were still wet, and when I forgot this and walked on them my boyfriend was distraught. —*Can't you see?* Can't you see how hard I'm working to get everything ready?

Did you know, did you know. If pregnancy were much more energy intensive, the body would begin eating itself alive.

—

A pair of socks with charcoal soles. A drawer of T-shirts with increasingly stretched seams. As I walked by a man smoking outside the post office, he held his breath; I heard the long exhale once I'd passed.

It is curious to me now that my central memory of these last days of pregnancy is not one of heaviness, but a counterintuitive lightness – as though I were dissipating, or being inwardly stripped. Sometimes, tiptoeing down to the bathroom at night, I was aware of a subtle moving of weights, of focus – as though some part of my inner balance system had or was about to shift. It seemed to run counter to the narratives available to me about what happened in motherhood, suggesting as it did not a tightening of attention, not a focusing inward, but a dispersal – a movement out.

Suddenly I wanted to clean everything. I wanted clean windows, a clear view. I filled the sink bowl in the kitchen and carried it outside, soap suds slopping over the concrete yard as I hauled myself onto a wooden stool and began sponging the dirty glass.

—Nesting already, are you? A neighbour laughed, appearing from his house. I felt wrong-footed. Caught investing in some needless preoccupation, some redundant busyness: mother, domestic, diminished.

But nests, I recalled, can take many forms. They're built above and below ground, and on cliff faces, and in the hollows of trees. They might be painstakingly woven, the work of many seasons, or nothing more complicated than a series of sticks dropped from mid-air and caught, pell-mell, in a tree.

Thinking this, I laughed. I scrubbed harder. So what if I was nesting? Who could tell, yet, the kind of nest I'd build? The dirt loosened, ran free, sliding down the windows and working its way beneath my fingernails. I was almost full term now and the baby was still upright, they still hadn't turned, despite my spending a good part of each evening practising headstands against the wall. Yet through the silent fact of this resistance, they had somehow become more real to me – more surely alive in their own right. I soaped again, and rinsed again, and the water and the dust and dirt ran into the drains and was washed away, as the baby jabbed and I gasped, while in the sky starlings appeared, crying out, careening.

II
BIRTH

The first time I heard my daughter cry I was strapped to an operating table, numbed from the waist down, and she was on the other side of the room, hidden behind a person or it might have been people in hospital gowns. Someone, some moments before, had whispered from behind my left ear that she was out (they didn't say 'born'), so I had known to listen for her. There was a lag, though, between this anonymous whisperer and the sound of her scream; a breathless wait in which – what? She gasped? Was suctioned? Her mouth, nose, throat and lungs struggled against the foreign substance into which she had just unceremoniously been dragged?

Grey whales, I've learned, emerge not into water but the air. The mother swims upside down, her flanks breaching the ocean's surface; her calf is born head first, skywards. Our first breath is deeper than the rest, and slower. The next are irregular, interrupted. By sixty minutes, the repeated intake/outtake has usually fallen into a pattern.

When it came, her sound, it was high and clear and the realest thing I've ever heard, and immensely far away. Moments ago, she'd been inside me; now and ever after, she was not.

I lay there, immobilised, teeth chattering insanely – a side effect of the anaesthetic. At my shoulder, my boyfriend reminded me to breathe. Again I waited, until finally I was handed her – soft, purpled and dressed in an overly large knitted hat. Days later, I would see this hat lying on a side and realise that in fact it was not large at all, indeed it was very small, and I would understand then that the thirty-eight-week emergency scan had not been wrong; that she was indeed very tiny, almost in fact too tiny – that something in the placenta's system of delivery had failed such that I had given birth not to a baby but a sparrow – a sparrow had come, been taken, 'out'.

She landed on my chest. Bony, feather-light, her limbs furled. In fact, to say that she was sparrow-like is inaccurate. It is too specific. She was simply all creature; all wild thing. Before language, before culture, before thought, confusion, longing, I saw now, we exist first as this: body. Need. Raw flesh. A deception too, though – she arrived with me *clean*.

I knew what newborn babies looked like from the telly. On hospital dramas I'd seen infants emerge blue and bloodied and covered in vernix – a thick, cheese-like substance made up of fatty glandular secretions and dead skin cells that works to form a moisture-retaining barrier in the last stages of pregnancy (yes, it's true – our cells have already begun dying before we've even been born). If left on a newborn's skin, vernix can continue to protect against dryness and infection; by delaying the clamping of the umbilical cord, more iron-rich blood is able to pass through to the infant from the placenta, which keeps their blood pressure stabilised as they take their first breaths. So who'd washed my daughter, who'd severed the cord, before I saw her? Who'd washed my daughter before me? And how was I to enter motherhood without some visual evidence of what the two of us had undergone? A friend of mine had her placenta made into a tincture; another cooked hers up in a frying pan. I meanwhile had only the fact of our flesh; the large dressing taped across my stomach; the plastic tubes extending outward from my chest, hand and urinary tract; the anaesthetic's slow retreat.

Among the circles that I tend to move in, these bodily substances and parts (the vernix, the cord, the placenta) are accorded value. They're markers of authenticity, of health; to perceive one's body as a source of inherent good is to feel that one is getting it right as a mum.

Some weeks earlier, I'd met the friend of the fried placenta in a cafe. Leonie's neon fingernails had tapped the plywood counter-top as she told me childbirth puts you more in touch with your animal side. —My animal side? I'd said. —You know, she said. —Your creature self. You become more intuitive, more primitive.

I'd cringed, and bitten my lip – sceptical, transfixed. As we took our coffees and found a seat by the window, Leonie told me about the birth of her son. She'd wanted a natural birth and she got candles, a birthing pool and a track of dissolvable stitches; they played recordings of whales singing and visualised water flowing into deep wide pools and it was wild, she said. It was awful, it was ecstatic. —There were murals painted on the walls of the labour room. I gave birth in the company of fucking Donald Duck! —How did you do it? I asked, meaning the birth, not her cartoon audience. —How did you stand it? —I moved inside, she told me. —I went really *deep*.

Here, it seemed, was another form of naturalness, different this time from the one co-opted to sell babygrows. The birth Leonie described was not calm and serene, but messy and uninhibited. It could be accessed not by making the right shopping choices, but via a more visceral encounter with bodily fluids and flesh; via (crucially, it seemed) the vagina, which was of course the precise component I later bypassed as I lay on the operating table, unable to feel below my waist.

If I was intrigued by the idea of a creatureliness that lay (obscured, ever present) within us, I was also wary of it. To be in touch with my animal side was, I construed, to be subsumed by the maternal instinct, by innate knowledge about what to do; to act unfettered by the mind. Yet what was the outcome of this line of thinking? That if the labour is difficult, if it doesn't go to plan, your body – you – can seem to be at fault for not being natural enough; that the experience of other species is short-handed, that their detail is lost.

Moving in, moving deeper, I wondered what I might find. And what did it say about Leonie and me that as we sat in the cafe that day, wiping steamed milk from our lips and laughing about a duck dressed in sailor's costume but not recordings of whale song, we policed the terms of our own experience, parcelling out what was natural, and what was not?

*

I never went into labour, so I never had the chance to find out for myself, in my body, what Leonie meant by her comment – what creaturely powers of determination and strength she'd wanted to prepare me for. Had I lived in a part of the world without access to free, safe obstetrical care, it's possible a 'natural' birth would have resulted in my daughter's death, or mine – a version of Nature omitted from the manuals I read, despite being entirely in keeping with our species' rather peculiar approach to birth. As it was, I was booked in for a C-section with two days' notice and the delivery was highly medicalised, time-limited and efficient. As it was, I witnessed an atmosphere in the operating theatre of quiet focus, then celebration. As it was, I don't know if the first thing my daughter felt of this world was the stainless steel of the obstetrician's tools, or her hands, or the air.

—

Some creatures release their ova indiscriminately into the world. Some (frogs, sea slugs, some insects perhaps) may feel nothing at all as they lay their eggs; for live-bearing reptiles, whose offspring emerge through their cloaca (a single orifice combining digestive, reproductive and urinary tracts), the sensation may be akin to having a shit.

At Leonie's suggestion, in the last weeks of pregnancy I'd signed up for a hypnobirthing course. I liked it. I liked the thought of being able to outmanoeuvre pain. On the last day we were given scented candles and a set of natural birthing affirmation cards to read every night before sleep. *I am relaxed and calm. My baby and I are safe. My baby is the perfect size for my body. My body and my baby work together as a team. I trust my maternal instincts, I know what to do.*

I'd enjoyed the class, but that evening, reading over the statements, my head clouded with counter-arguments. Attending to anxiety can be important, can it not? Also, not all bodies are safe. Not all foetuses are 'perfectly sized', and to consider them team players is surely to be missing the point.

Human babies have big brains and lots of body fat, which involves an enormous outlay of resources on the part of the gestating mother. In her book *Growing up Human: The Evolution of Childhood,* bio-archaeologist Brenna Hassett argues that pregnancy conditions such as pre-eclampsia and gestational diabetes may result from our uniquely baby-led pregnancy; the foetus isn't working *with* the mother so much as taking what it needs to sustain itself.

While other primates tend to give birth in under two hours, the average human labour extends to nearly nine, and the need for assistance is almost universal across cultures. For humans, a 'natural' birth is often a difficult and painful one – and the matter of sizing is a moot point. Somewhere between 3 and 6 per cent of human deliveries are obstructed (the rate varies, depending where you live), most commonly due to a mismatch between the infant's head and the mother's pelvis, which tilts, meaning the baby must rotate during its passage through the birth canal. Ultimately, the manoeuvre is complicated, the head only *just* fits, and some babies don't fit at all.

A 'natural' birth was, as I understood it, smooth, instinctive and unproblematic; it was supposed to feel like flow, to sound like waterfalls and whales. Yet this perhaps betrayed a rather faulty conception of what Nature is actually like – of what coming to motherhood is *like* – for humans, as well as other species.

Until as late as the 1980s many scientists remained unsure as to whether other species were able to feel pain, since such a subjective psychological experience, combining both sensory and emotional aspects, implies a sentience that, most believed, animals lacked. We now know that lots of species, including many invertebrates, *do* experience pain, and possibly other emotions too. It's a shift that has necessitated a greater recognition of the fact that difficult labours aren't a uniquely human problem; that birth in the animal kingdom can be far from easy.

Horses sometimes sweat while giving birth; llamas and alpacas bellow or hum as they do when injured; chickens' facial expressions change during egg-laying (though it's difficult to say what they're feeling). For some, labour is more extreme. The spotted hyena gives birth through her clitoris, or 'pseudopenis' – an extendible phallic structure with a false scrotum and testes and a birth canal measuring just one inch in diameter. The porcupine's offspring, known as 'porcupettes', emerge fully quilled (the spikes don't harden until they're exposed to air, but they can make for complicated breech births). The eggs belonging to the flightless kiwi bird are so large that laying one is proportionally equivalent to a human birthing a four-year-old child. The birth of an infant squirrel monkey creates a wincingly tight squeeze; in one captive population, stillbirths were reported in 50 per cent of all deliveries.

Such births are not straightforward or smooth. They're tricky, they contain particularities – they're borne of bodies that are at once strange and full of detail. I think, too, that they hint at what is perhaps a more troubling realisation: that birth and labour are life-and-death stuff; that they can be risky and brutal. With this, they also trouble a surprisingly persistent notion of the 'natural' animal mother.

In *Bitch: A Revolutionary Guide to Sex, Evolution and the Female Animal*, zoologist Lucy Cooke explores how from the time of Darwin and for much of the last century, scientific knowledge was profoundly influenced by a set of gendered assumptions about what was natural, leading to a series of misconceptions that extended far beyond the mechanics of birth. Dismissed as a simple and mostly homogenous group, female animals were largely considered to be passive and unchanging, little more than (unfeeling) baby-making machines – the polar opposite of their more powerful, more promiscuous and more aggressive male counterparts. As a result, they went underresearched, their experience overlooked, with any anomalies tending to be explained away (Cooke gives the example of ornithologists who described aggressive encounters between female pinyon jays as the 'avian equivalent of PMS'; of zoologists looking the other way while female lions mated 'scores of times a day during oestrus with multiple males'). Through time, a data gap opened up, and the myth of the automatic, unvarying animal mother became, in Cooke's words, a 'self-fulfilling prophecy'. In a culture inclined to draw inferences the other way, from animal behaviour back to human society, such a narrow conception of female animals served to further reinforce those same gendered assumptions about what was natural and normal for our own species.

Smooth, simple, passive, unchanging. I recognised these attributes. They were present, in an implicit way, in the images designed to sell me things; perhaps they were there even in my own imagining of how, beyond childbirth, I would be – how I *should* be. Yet in recent decades, in zoology and evolutionary biology at least, the emphasis has shifted. Many have been working hard to fill the gaps left by past biases, and as a consequence, scientific understanding has thickened and complexified. Some of the findings are surprising; many disturb or upend ideas that had been in place for generations. Female animals no longer appear as unvarying, or as simple, as they once did. Primed as we are to draw comparisons between ourselves and other species, this developing research challenges us to think again about mothering, and about other animals – not just what separates us, but how we're joined.

Most primates seek seclusion during labour, but among those that form pair bonds, fathers are often present at birth and typically take an active interest in the process. Though birth assistance was once considered a defining feature of our species, we now know that while it is unusual among other primates, it certainly occurs. Captive orangutans have been observed helping partners during delivery; common marmoset fathers act as midwives, grooming and licking newborns – a role that may have evolved as a result of the high energetic cost of birth for the mother, whose twin babies make up 25 per cent of her total body weight.

For some species, having other group members present during birth may offer protection for newborns who would otherwise be at risk of attack. For others, such as the black-and-white snub-nosed monkey, more experienced females appear to provide practical assistance to labouring females. Such interventions may be more likely to occur at times of crisis – as when a hamadryas baboon reached out to catch an infant being born over the edge of a cliff, or when an experienced *Rhinopithecus* monkey dragged a new mother's infant out of her birth canal and broke her amniotic sac.

For bonobos, our closest relatives alongside chimpanzees, females are self-sufficient in accomplishing delivery, yet they seem to prefer to give birth in company. One study describes how a group of female attendants, some of whom were mothers themselves, were seen guarding a labouring individual from approaching males, swatting flies away from her exposed genitals and even trying to catch the baby as it emerged. These findings have implications for how we understand birth in our own species. It's possible, the researchers note, that forms of midwifery were present before our own evolution made the practice obligatory for most people; that it emerged initially not out of physical necessity, but rather the capacity of our deep ancestors to form strong bonds and work together.

On the hospital ward I lay in bed, the baby upon me, everything tinted in spearmint and washed out by fluorescent bulbs – weird, ecstatic, wired. Lightly, she worked at my chest, her soft mouth searching for a latch, whereupon came the next bombshell of her life: my determinedly inverted nipples. Determinedly, because I had in pregnancy tried various strategies to evert them, experimenting with little suction pumps and silicone doughnuts, yet I'd had no success. If she was to feed, she would first have to evert them herself.

The midwives were encouraging, if cautiously so. One of them wheeled out an electric breast pump and proceeded to stand with an air of distant scrutiny over by the curtain. Yet the baby continued to suckle, working at the puckered skin with apparently untroubled resolve. (Is this what we possess, I wondered, before anxiety? Before the experience, and so the idea, of failure?)

I was barely holding her – more, I was focused on remaining very still beneath her, not wanting to disrupt her work. Like this, I got a better look at her hair, her fists, her near-translucent eyelids. Gingerly, I nudged her, closer in. My boyfriend, watching from the other end of the bed, touched my toes. This was distracting. I wondered about asking him to get off. How to square my need to concentrate wholly on what was happening at my breast, her mouth, with his need to be a part of it?

Suddenly, the midwife cried —She's feeding!

I had thought at first that she was sparrow-like, but perhaps more accurately she was reptilian, I considered on that first night. Her bony head and those angular, shrunken-looking limbs; the looseness of her skin around the joints. Yes, there was something lizard-like about her – something ancient. Was it not true that, *in utero*, we reveal traces of other, older selves? In the second week of pregnancy, a yolk sac would have formed inside my womb; an evolutionary echo. In the fourth week, she would have been indistinguishable from a fish embryo at the same stage; in the seventh, her hands would have developed a particular spread of muscles common to reptiles (these muscles would have degenerated as she grew, until at birth they'd disappeared completely).

She slept and slept and every three hours we woke her, as we had been told to do, and I attempted to keep her awake and latched – tickling her toes, pulling off her vest, with no real sense as to whether she was feeding or not; and then after half an hour or so she would be asleep again, whereupon we would return her to the cot, and my boyfriend would lie cocooned in a sleeping bag on the floor, as I, unable to sleep, listened to the sounds of the hospital at night – trolleys; footsteps; howls.

—

Remember the study of captive squirrel monkeys, where stillbirths were recorded in half of all deliveries? Once an infant's shoulders are freed, he heaves himself out using his arms and hands – a level of independence not possessed by our own species, who emerge startlingly helpless, though it's possible our unusually intense grip in the first hours of life may offer a different signpost, to another part of our ancestral past.

Reflecting on these remnants, these traces of evolutionary history, it's easy to believe I don't have just one animal side. Perhaps instead I have lots of them. Some three billion years ago, we were single-celled organisms. Later, we were sponge-like creatures, then worm-like ones; then fish, amphibians, reptiles, early mammals, early primates, apes – a single thread in a densely branching evolutionary story. Thinking this, I'm reminded of the claim that amniotic fluid has the same salinity as seawater; a neat signature of our watery past. Not true, quite – but the concept, I realise, is not as far off as I'd thought. Most amphibians today still return to water to breed; twenty thousand unfertilised eggs might spring from the female bullfrog, as the male atop her back grasps her with his forelegs and the rough skin of his thumbs to release his sperm. From some amphibians came early reptiles, and the arrival of the hardened eggshell. Fed by a yolk sac, these embryos developed in a pool of contained liquid – now, they could be laid and incubated away from water. From here, so the story goes, emerged viviparity, or live birth, where the embryo gestates inside a fluid-filled amniotic sac. For viviparous species, the pond is inside the body – inside the womb.

Begin thinking like this, in terms of evolutionary time, and motherhood soon sheds its associations with stasis. Instead, it appears unsettled, in process – full of proliferating branches and constant balancing acts. A move from external to internal fertilisation increases the chance of offspring survival, and allows females to be choosier about mating partners (a surprisingly active role for the purportedly passive female), but it's more demanding in terms of bodily resources. A move from egg-laying to live birth is more energetically demanding again, and leaves the gestating parent vulnerable to predators – yet it also increases the chance of offspring surviving into adulthood, and so allows for the possibility of smaller litters.

Tracing the branches, automatically I begin assigning species to categories – crocodile: oviparous; rattlesnake: viviparous – but quickly things get complicated. In fact the rattlesnake's young develop inside eggs that are held and hatch inside her body; lacking a placenta, yet giving birth to live young, she is termed in the literature as *ovo*viviparous. So what of the snake whose eggs hatch within twenty-four hours of laying, having gestated inside her body for many months? Or the seahorse who deposits some 1,500 eggs into the male's remarkably womb-like pouch, for them to emerge fully formed some weeks later?

There are categories within categories, and some back-and-forth between them. The Australian three-toed skink, a bronzy snake-like lizard with a petrol-like sheen, is capable of laying eggs and birthing live young simultaneously, within the same litter. In evolutionary terms she exists in a woozy in-between state, transitioning either from egg-laying to live birth, or back again.

If I had pictured evolution unfolding in a linear way, a steady chug-chug of advancement and progress, the skink suggests something different. A stranger, more variable story of blurring categories, shifting states. And stories matter – they shape what we see when we look at the world, how we make sense of ourselves. Which is why a recent paper by researchers at Bristol and Nanjing universities arguing that viviparity may actually have occurred *before* the arrival of the hardened shell is so striking. The team revealed evidence of viviparity and extended embryo retention in early reptiles; a finding that calls into question the linear view that what enabled early reptiles' transition onto land was the emergence of the hardened shell (the pond inside the egg), and explodes the notion of the simple, unvarying animal mother.

Extended embryo retention is common among snakes and lizards alive today. It occurs when embryos are held by the body for a variable and sometimes extended period, helping enzymes to remain active if the environment is unusually cold, or as an added protection against disease or predators. As with the rattlesnake, eggs are sometimes retained until and even beyond the point of hatching, meaning that usually oviparous females actually end up bearing live young. Following this logic, it's possible that the critical factor determining reptiles' movement onto land lay not in an aspect of the egg's physical architecture, but the ability of females to adjust the timing of birth, retaining their embryos for longer periods – a characteristic that is skilled and dynamic, rather than static or mechanical, and one that we might assume will be increasingly important in the uncertain weather systems of the future.

It was one o'clock in the morning, and in the bed next to mine, a woman on a video call was describing the three rounds of IVF she and her partner had undergone. —They wash them, she said, speaking of her partner's sperm. —They spin them really fast, and all the dud ones fall down, and the ones that are left are really strong.

Perhaps it was the lateness of the hour. Perhaps it was the effect of the painkillers, or the influx of hormones, or the surfeit of new life amid all the rigidity of the ward. Whatever, the conversation became stranger. —Soon, the woman's friend was saying —*soon*, they'll be able to make babies without using eggs or sperm. Skin cells, hair follicles, that sort of thing. You could have a baby with Keanu Reeves. You could reach out and scratch him, and make a baby. —Like something out of sci-fi, isn't it? said the woman next to me. —Like aliens.

Or, I thought, *like worms*. The hermaphroditic flatworm might reproduce by injecting their sperm into the skin of another, then continuing on their way; they might fertilise their eggs with sperm from their own body; they might also split off a piece of themselves, turning into two.

I couldn't sleep, and neither, clearly, could my neighbour. My boyfriend had wheeled the baby off down the corridor to let me rest, but without her there beside me I felt even weirder than before. Was this my instinct, kicking in? Groggy and confused, I shuffled along the corridors looking for them – now hearing a baby's cry; now convinced that it was hers.

They were in the TV room, and she was fast asleep. The sound on the telly was turned right down, and there was music playing softly through a Bluetooth speaker. On a chair in the corner, a man in slippers and a tracksuit was snoring loudly, mouth agape. I glanced at the headlines flashing as subtitles across the screen. Logging in the Amazon. Jeff Bezos' space flight. A virus, spreading fast.

Outside the window, a wind – tree branches rattling, just visible in the light cast from the car park. I sat down beside the transparent sides of the hospital cot. Perhaps after all she *was* bird-like, I reflected then. Seeing how she coiled, her muscles held by the memory of confinement, I found I wanted to re-encase her; to make of myself a force field, a shell. A vestigial memory, perhaps, of an earlier time – or else a riff on how I'd learned to see myself in pregnancy. A barrier, a membrane, an impossible responsibility.

—

To think of evolution is to think of branches, but I don't imagine them above ground. Instead, I see roots – intricate, entangled, subterranean. A different branch and it would have looked like this: her, winged and hairless, inside a mineralised shell.

A bird's egg, then. Here, too, a site of balancing acts, from the size of the yolk to the timing of when it is laid. An egg laid too early has less chance of survival; too late, and the mother is unable to meet the oxygen demands of the growing embryo inside. Yolk size determines how many nutrients are passed to the unborn chick, and thus how far they're able to develop: owls lay eggs with a yolk occupying only a quarter of the shell, and chicks emerge needy; for geese the figure is closer to 40 per cent, and chicks are able to walk and feed themselves soon after hatching.

Was it our second night in hospital that she began screaming? Red raw bundle of panic-stricken, indecipherable need, which we passed back and forth between us, wondering what in God's name to do.

The foreverness of those cries. Their interminableness. The feeling of utter porousness, and unpreparedness, in the face of them. *How on earth to make them stop?* Outwardly, I suppose I appeared calm. I hushed, I soothed, while all the time a storm was under way inside me – I couldn't remember ever having felt so *open*.

A midwife passed by the curtain, apparently unfazed by the chaos being unleashed in our arms. We held, we cooed, until finally, just as suddenly as she'd begun, the baby calmed.

Did you know, I whispered to her in the early hours, as she slept and as he slept and as my heart rate slowly returned to normal. *Did you know*, I whispered, *about the malleefowl?* The malleefowl, a stocky ground-nesting bird with flecked, dust-coloured wings, lays an egg so rich, the yolk occupies over half of its shell. The nest is a purpose-built mound of dead leaves, which incubates the egg through the heat generated by the process of decomposition.

To lay one's egg in a pile of dead leaves; to bury it in decomposition, in process, in the dark. *That is how they care*, I whispered as she lay, feather-light and sleeping, beneath the hospital bulbs – not knowing then what I meant by it, nor why it moved me so. *That is how they care.*

———

Now and then, at allotted times, midwives came and ticked things on a spreadsheet and handed me little plastic cups in which rattled more painkillers. In the booth opposite, a Polish woman, just out of surgery, wanted to go home. —You can't go yet, the midwife told her. —You have to get the all-clear from the consultant. —When he's coming? she asked. —This afternoon, the midwife said. —I have three children, the woman snapped. —Who's going to look after them? They offered her painkillers too, but she refused them.

With free hot meals, fresh sheets and around-the-clock midwifery support, I was in no hurry to get home, but in any case it wasn't on the agenda because our baby kept turning yellow. Blood samples were taken for extra tests; I was hitched up to the electric breast pump and told to keep a check on how much thick, creamy colostrum, or 'first milk', I was able to produce. 10.5ml, 10ml, 8ml . . . the measure seemed to be going down.

Other parents I knew by voices only. In the far corner, a quiet, mournful woman who barely exchanged a word with her partner, except to tell him to pass her things (mascara; a babygrow; her phone). Beside them, a younger couple, the father so full of nervous energy that he kept rushing out to the reception desk and back, brimming with questions about everything from breast milk to parking permits. Over by the door, two women spoke in hushed voices and made frequent visits to the coffee shop downstairs, filling the ward intermittently with the smell of croissants.

These people came and went and still we remained, cocooned in our spearmint booth. On the afternoon of the third day, as we shared my hospital-issue meal of curry and rice, my boyfriend, looking at his phone, frowned. —What is it? I said. —I think I need to get to a shop, he replied, and left the ward soon after. But it was too late. By the time he reached them, the supermarket shelves were already emptied of nappies and painkillers, loo roll and baby formula. If in pregnancy I'd led myself to believe that I was developing an instinct for detecting threats present in my environment, I'd managed to overlook a big one. The world had changed in the few days we'd been away, and the virus about which newsreaders had been warning had turned pandemic.

As he went home for a bath and a change of clothes, I sat glued to the bed, unable to put the baby down. Each time I tried, she screamed until I picked her up again. Soon I needed to pee. Bewildered at this sudden impotence, I pressed the buzzer, called a midwife. A head appeared around the curtain. —I need the toilet, I said, pathetically, uselessly, and she took the small bundle as I hobbled into the only toilet on the ward – a small, overheated cubicle thick with the stench of blood and piss and whatever else it was that had leaked from us and not yet been cleaned away.

Back outside, the midwife stood serenely, and the baby's eyes were open. A new kind of gratitude, then, towards another woman – her hands, her arms, her warm and willing presence. I glanced down at my T-shirt. It was drenched with milk.

———

Some years ago, a friend of mine began compiling a list of births that had gone viral. This was a compendium of animal births, not human ones, and she used to reel it off on request, this collection of live-streamed labours. There was Zoom the African elephant, born at a safari park in Mexico; Kipenzi the giraffe, born at Dallas zoo; April the giraffe, whose labour at the Animal Adventure Park in Harpursville was for a time the second most-watched video in YouTube history . . . the list went on.

Ana used to work for a conservation charity, where she spent much of her time trying to convince people to care about the plight of species that live many thousands of miles from them. Viral births were fascinating to her because they achieved what she'd struggled to: mass appeal, mass *care* – sometimes even mass mobilisation. April the giraffe's labour prompted an online fundraising campaign, Toys 'R' Us sponsorship and a clothing line – unwittingly, she raised thousands of dollars for conservation projects worldwide.

Stories like this amused Ana, and left her perplexed. Culturally, there exists a hierarchy of animals – a hierarchy of how much we care. Furry, rotund animals are more popular than slimy ones, I remember her saying. Big eyes are more popular than smaller ones. And when it comes to mothers, the big draws are mammalian, milk-giving, known for high levels of infant care – 'good' mothers from a 'good' Nature, where all is abundant and in its place. Which is to say that the gendered stereotypes among the researchers of the past are not limited to the researchers of the past; that they linger in our gaze, and in our expectations of other animals, and what we want them to tell us, and how we want them to make us feel.

Once, Ana told me about the animals that hadn't made the list. A Komodo dragon at Denver Zoo who underwent emergency surgery after a scan revealed a clutch of eggs had fallen into her abdominal cavity and burst; a giant panda at Hong Kong Zoo who, following a four-year effort at insemination, reabsorbed her foetus. Another giant panda, Tian Tian, this time at Edinburgh Zoo, whose live stream suddenly went dark a few days before she was due to give birth, causing consternation among her waiting fans. Following a freedom-of-information request, the zoo's spokesperson gave a carefully worded statement: 'Her hormone levels and behaviour have returned to normal as the breeding cycle ends for this year'. She had, in other words, aborted the pregnancy, her body sealing off the placenta's food and blood supply such that the organ was no longer capable of keeping the foetus alive.

For a culture that has historically viewed female animals as synonymous with mothers, and mothers as synonymous with Nature, there is something discomfiting about Tian Tian's story. It goes against some of our most consoling narratives: that Nature always endures, that a mother's purest instinct is to protect and care for her young. The insight that a panda may abort her pregnancy, and that the move may be both natural and adaptive, invokes the role of another, further factor: that of outside circumstance, or environment.

'The goal of motherhood isn't to nurture babies indiscriminately,' writes Lucy Cooke, 'but for a female to invest her limited energy in creating the maximum number of offspring that survive long enough to reproduce themselves.' Animals are fundamentally concerned with their own survival; they must find food and shelter, and escape predation, if they're to pass on their genes to the next generation. Giving birth and raising young puts them at greater risk – if parenthood comes at the wrong moment, both mother and offspring will perish.

In reptiles, extended embryo retention evolved to factor for exactly this – though the Komodo dragon at Denver Zoo held on to her eggs for so long, they broke the oviduct wall and fell through it. In placental mammals, gestation length is usually fairly fixed (it rarely varies by more than 3 per cent), but for a few species it's surprisingly flexible. Bears, bats, deer and most seals are all able to suspend gestational development through a phenomenon known as the 'embryonic diapause', which can last anywhere between a few days and many months. Meerkat pregnancies vary by up to 25 per cent when females are under stress. Mink mate in March and hold on to their embryos until after the spring equinox, when outside temperatures are usually more favourable for raising offspring. Kangaroos mate soon after giving birth, and the onset of lactation triggers a suspension of the new pregnancy until the suckling infant is fully weaned, meaning that an adult female might be carrying three joeys at any one time (the first almost independent but still suckling, the second immature and living inside her pouch, the third a suspended pregnancy). If this sounds like hard work, it's also a marker of the kangaroo's evolutionary resilience; chased by a predator, she may abandon the bigger joey – he won't survive, but the cessation of suckling triggers the suspended pregnancy to begin again, meaning another infant will soon be on the way.

For polar bears, bringing a pregnancy to term is contingent on internal as well as external cues. They mate in late spring as Arctic temperatures begin to rise, and the female will hold her tiny embryo in suspended animation until autumn, when, if she is strong enough, it implants into the uterine wall. For the polar bear, motherhood is an act of extreme endurance; she gives birth in winter and won't leave her snow cave until the following spring, during which time she won't eat or drink – her cubs will continue nursing until they're at least twenty months old, and will stay with her for almost three years. If she hasn't gained sufficient weight during the summer of her suspended pregnancy, it will not continue beyond the diapause.

In some species, pregnancy loss has a social element. Among gelada baboons, who live in complex social groups, 80 per cent of pregnancies are terminated in the weeks after a new dominant male arrives – a move that avoids wasting unnecessary resources, since the incoming male will kill any offspring that don't belong to him. (Learning this, I find myself asking if it is more than that, too – if a species, feeling more than we know, might evolve an adaptive defence against certain kinds of pain.)

It sounds like agency, does it not? Or at least a more contingent, fragile kind of balance. A counterpoint to the image of the passive, unchanging mother whose task is to churn out as many offspring as possible. I'd like to know what it would take to hear the stories these creatures have to tell; what they might teach us about animal mothering, about mothering naturally, if we could only understand. I've tried imagining, sometimes, what happened in those moments after the live stream at Edinburgh Zoo went dark. What viewers might have glimpsed in Tian Tian's returning gaze, had the camera kept running. Perhaps nothing – perhaps only our own imaginings about what it is that other animals feel. Yet her body, its action, seems to present us with a statement: *not like this*, she asserts. *Not like this.*

In the forty-eight hours between arriving back from the emergency scan and returning to hospital for the C-section, I'd sat at the table in our front room and looked up Ana's live-streamed labours. The air was cool; the lane was quiet. There was a sense of stillness to the house, if not completion. Had we put our minds to the right jobs, the right areas for improvement? Were we as ready as we should be? I followed the grain of the table with my fingernail, curled my toes inside my socks. Was it enough – was it all OK and enough?

Leaves skittered along the road outside and I felt again that internal shifting of weights, of attention. Yes, a shifting, a recalibration, too big to look at head-on.

I was treating this as a project. I was treating it as research, which both explained and legitimised the compulsive gathering of information, the collecting up of stuff. To begin a process of research was to permit myself the mess, the untidy things I'd dragged below ground, the disparate themes that didn't connect and wouldn't connect unless I stayed with them, as I am staying with them still, watching and waiting for a sense to form.

Here's an image you won't forget easily: giraffes give birth standing up. The calf plummets six feet to the ground, causing the stretch-stretch-*snap* of the umbilical cord that had held them together.

In the run-up to April the giraffe's labour, her fans wanted regular updates. Newspapers covered the story. The live stream was briefly pulled off-air when animal activists complained that it violated YouTube's policy on nudity and sexual content, but was later restored following thousands more complaints from viewers who wanted to keep it there. Now the audience numbered over thirty million, and people were asking about the father. Zoo officials explained that Oliver had enjoyed some yard time with April in the days leading up to the birth, but that the pair were currently being kept in separate stalls. '[His] rambunctious play [. . .] could have negative effects,' they reported. And then, bafflingly, bewilderingly, 'boys will be boys'.

Those last words sprang from the screen. What 'boyish' behaviour was being referred to here – what play was being legitimised, what 'negative effects' were at issue? And what did it mean that a captive stall, scrutinised by thirty million, had become synonymous with a new mother's protection?

I scoured the internet, suddenly intent on ascertaining the details omitted, the version of masculinity implied – the parallels being drawn between the males of this strange, lonely-looking species and our own. I circled animal blogs and websites, each of which seemed to draw from the same limited pool of 'fun facts'. I learned that the male giraffe drinks the female's urine as a precursor to mating (it's a way of detecting the presence of oestrus, which will tell him if she's fertile); that in the wild, pregnant females return to the same calving area for successive births; that they like to give birth alone, in seclusion – to go through labour unobserved.

On the fifth day in hospital, the baby pinked. The yellow tinge was gone, she'd passed the necessary tests and our bed was needed – we were released, sent home.

It was dark by the time our paperwork was done and we were allowed to leave. The baby was bundled into blankets and her silly oversized hat, strapped into the car seat we'd inherited from a friend. I hobbled along the ward, my stomach twinging sharply inside a fresh dressing as a hospital caretaker screwed a bottle of hand sanitiser to the wall. At the reception desk, the midwife who'd held the baby two nights before pressed a paper bag into my hands. —I'm not supposed to do this, she said, and put a finger to her lips. I peered inside: three packets of painkillers. Tears sprang. She winked. —Good luck.

The baby slept. As we drove onto the city bypass, I willed my body, unpredictable at the best of times, into action – into production. *Milk*, I urged my breasts. *Milk, milk.*

The roads were almost empty. We streamed out of the city, into the hills. I found myself thinking of the day I passed my driving test, when I took my mum's car out for the first time. I was so aware of the *space* around me – and yet it was a peculiar feeling, since alongside this spaciousness was a sense that things were simultaneously very close. Pavement kerbs, buildings, lamp posts – all were suddenly within reach, and perilously so, since the thing about being responsible for a moving vehicle is that one could crash at any time.

I was sitting in the back, beside the baby. I could not decide whether to gaze at her or at the night, which sparked with electricity. *Are you feeling connected with the baby?* I remembered the midwife asking, early on in the pregnancy, and how out of place that question had seemed at the time. Now, I had an answer. *Yes*, I wanted to tell her. *Yes, yes.* But the connection was not as I'd imagined it. The connection was snappy; it shot through everything. The connection was wild, it was immense.

III

ANIMAL FORMULAS

In a free breastfeeding class I attended towards the end of pregnancy, a nurse held up a small polythene bag filled with coloured plastic beads and explained that they represented all the ingredients to be found in breast milk. There followed a moment of confusion among the group of eight or nine women arranged in a neat crescent shape around her. We were accustomed, I suppose, to associating long ingredients lists with manufactured foods – ready meals, cheap chocolate, cakes with long shelf lives. We were accustomed to gauging our pregnant bodies using fruit and vegetable scales. But these, the nurse emphasised, giving the bag a shake, were good ingredients. These were *health-giving* ones. She reeled off a list, which I promptly forgot, instead committing to memory only that image of the plastic bag, its colourful concoction; the way it rattled when she shook it.

She reminded me of a primary school teacher. Perhaps she had kids that age. —The composition of your milk will change depending on the time of day, she was saying —and on the baby's stage of development, and in response to a lot of other factors, from your own past experience and disease history to your baby's changing health. (The illnesses to which a breastfeeding mother has been exposed affect the make-up of her milk, I later learned – her body effectively builds an infant physiology based on its expected life exposures. Similarly, when a baby becomes sick, his saliva seems to signal which antibodies are needed to help combat the infection; as he starts exploring the world, putting things in his mouth, her body responds by producing milk with more bacteria-fighting enzymes.)

The nurse paused a moment, letting that settle. I wondered if she was working from a script. Perhaps she'd done this class so many times she'd memorised it. I glanced around at the rest of the group, most of whom were younger than me, and at their bumps and breasts, all of which were larger than mine. The nurse continued. I fidgeted. The chair was uncomfortable – one of those bright, stackable things that looks like it's been ergonomically moulded to fit the shape of your body, when in fact your body must mould to fit it.

The truth was that in relation to my own apparatus, all this felt rather conjectural. If at night I'd dreamed of milky abundance and overflow, nothing had so far appeared from my swollen and increasingly tender breasts but a hard and pus-like crust. Which lent a peculiar edge to the message implicit that afternoon that breastfeeding is a matter of choice; that the only thing standing between a mother and her ability to nurse is her belief in breast milk's superior status.

Geared as it was towards persuading recalcitrants to the cause, the class had nonetheless attracted the already converted. As we moved to question time it became clear that none of us was in any doubt that breast milk was good for babies, and for mothers too. We wanted to breastfeed; we were motivated. We imagined it as the purest, most natural form of feeding; a gentle, intimate kind of bond. And yet, quietly perhaps, we also doubted ourselves – we feared failing, feared this new and unfamiliar labour, this further recourse to the body, to *our* bodies. So, we'd come wanting to arm ourselves with information and equipment: what kind of steriliser, one woman asked; and what kind of pump, asked another. And how to tell if your baby is hungry, or full, and if they're properly attached, or if they're comfortable, and what if we're not? The feeling nagging, as we put up our hands, that these were perhaps not quite the right questions; that they were not reaching through to whatever it was we sought from this small and carefully turned-out woman – her bag of collected beads, and her air of calm and confident assurance.

At the end of the class the nurse handed out pamphlets bordered with cheery cartoon animals (large eyes, round bellies, big smiles). On the back was the number of a helpline, which we were encouraged to call if we struggled initially. But we shouldn't be put off, she said, if it was difficult at first. With a bit of practice, it would soon be second nature.

—

Inside the mammary gland is an inverse tree; a river of multiplying tributaries. I seem unable to describe it without reaching for natural metaphors. The tree (the river) reaches inward, towards the chest cavity, its branches proliferating with increasing distance from the skin. At the tips of these branches are 'lobules', each of which holds a series of hollow sacs. Here, during breastfeeding, milk is produced and delivered back through the tree structure to the nipple.

The mammary gland develops in stages, first in puberty and later, if a woman becomes pregnant, in preparation for childbirth. These maturational processes happen as a cascade of complex hormonal changes, the body readying itself in waves that through pregnancy I experienced as felt emotions: despair, fatigue, bliss, blind panic. It was odder than I expected, the changes more internal and less easily describable to anyone, least of all myself.

Still, I wonder now what I might have asked at the class that day, had I the words for it then. *What does it mean to be the source of this substance that already is causing me (my breasts) to become warmer, tougher, fuller? What if I lack the syphon for it? What if it won't come out?*

I shuffle the sentences like parts of a puzzle. I want a phrase other than 'come out' – a word other than 'express', other than 'drain'. *Extract, draw off, empty.* Each one, I notice, places the action on the part of the suckling infant. None describes the peculiar busyness of breast milk, nor how it might shoot from the breast unbidden, or soak through jumpers, or puddle in sheets as we sleep. The image of the passive, static mother runs deep.

—Did you know, I'd asked my boyfriend, the morning after the class, as I slipped a pair of suction cups inside my bra.
—Did you know, a seal has retractable nipples? They tuck

them in when not in use, probably to cut down on swim drag and as extra protection against the cold.

We were standing in our socks and pants, our bodies pallid and (I sometimes thought, now) aged-looking in the grey and early light. —The mammary organs of whales and other cetaceans are buried too, I added—in a fold of skin on either side of another, larger fold containing their genitals.

Without a language to describe what was going on inside me, the animals seemed to rise up, filling the spaces words wouldn't go with the strangeness of their bodies.

—Humpback calves lack lips, I might have added then. They latch by curling their tongues into rounded troughs, creating suction and a channel for milk to flow through. Like this the tongue turns rough and raw as the mother swims, following migration routes that might span thousands of kilometres. As the calf drinks, bouncing and blistering, the combined exertion of swimming and lactation results in a marked depletion of the mother's body condition.

My boyfriend, perhaps, nodded his head. Perhaps mentioned I'd mentioned this before. Perhaps suggested I write it down, make it into something, rather than chattering to myself.

Bleed. A verb containing both sides, since it can mean either to lose, or take from.

The platypus, I've learned, never evolved nipples at all. Instead, milk oozes from ducts in her mammary glands and collects in grooves on her skin, then drips from tufts of fur.

The opossum has thirteen nipples. Her offspring emerge blind, deaf and in possession of a disproportionately large set of claws, which they use to haul themselves over their mother's body to her pouch, where if they're to survive, they must latch for two months straight. During this period, the nipple swells in the infant's mouth, reaching down his throat and extending up to thirty-five times in length. Like this, the two (one distended, the other engorged) create a tether so tight that trying to separate them can result in tears to either nipple or lips, or both.

When five days after the birth we arrived home, the house was not as it had been. We got back in the dark, and after all the bright lighting of the ward, the double-glazed windows and overheated corridors, the house felt draughty and closer to the elements than I remembered. The chill air was suddenly too cold; the distance between my arms and the concrete yard, the kitchen tiles, was suddenly perilous.

A neighbour had come by to turn on the heating ahead of our arrival, and as an extra thought she'd slipped a hot water bottle into our bed. But the cap wasn't fastened properly and the bottle had leaked – the sheets, when we got to them, were soaking. We slept that night under moth-bitten blankets, on an air mattress over wooden floorboards; the baby in her cot beside us.

'Slept' is not, of course, the word for it. The baby screamed. She screamed and screamed. A midwife later reassured us, truthfully or not, that this is normal the first night a baby is home; they sense the changed environment. From the dimness of the womb to a bright hospital ward; from hospital to home. There are no street lamps on our road, and the night presses in – a dark without edges, cavernously big. Was this what she sensed – what bothered her?

I hadn't a clue. More pressing was the now unsupervised task of feeding her. And so I followed the regime relayed to me at the hospital, pulling her to me at three-hour intervals, holding her quietly, unknowingly, under low light. Like this, I heard cats fighting below the window; foxes mating in the street. A milk van passing, just before dawn – a freight I hadn't known existed here, and a sound that, I realised, I associated simply and straightforwardly with early childhood.

—

Plato believed the uterus was a creature, wandering through the female body; Galen thought it had horns. In Western culture, the association between women and animals goes back to ancient times, and among all the organs in a woman's body, the reproductive ones were considered the most animal-like. In her seminal work, *The Death of Nature*, ecofeminist philosopher and historian of science Carolyn Merchant describes how, in Ancient Greece, the feminine identity was inseparable not just from conceptions of animals but from Nature itself – the two entwined in a dual image that contained a heady mix of forces, both nurturing and powerful;

both unpredictable and magnificent. It was an image that did not sit easily, many centuries later, with those who wanted to reorganise society along scientific principles. At the core of the Scientific Revolution stood the idea of Nature as an economic resource – a passive, manageable system that could be put in service to men. For Merchant, this conceptual shift profoundly altered our relationship with the natural world, permitting man's dominance over living things, and legitimising the subjugation of any group deemed closer to animals, or what is primitive.

Mother, animal; mother, Nature. If the association is an ancient one, its significance changes depending on the cultural attitudes of the time, and how we arrange the world in our minds. As a child, I learned to arrange animals into a series of categories: insects, fish, amphibians, reptiles, birds, mammals – a list I understood in a linear sense, with simple organisms at one end and more complex creatures at the other (humans were at the far end, representing a kind of finish line). The categories are based on a taxonomic system formulated in the eighteenth century by Carl Linnaeus, a Swedish zoologist and physician who, after some initial indecision, coined the term *Mammalia* – meaning, literally, 'of the breast' – to describe the class of animals to which we, humans, belong.

It's a curious choice of term, and a subtly important one. *Mammalia* is derived from the Latin *mammae*, a word that refers as much to the organ's milk-secreting aspects as to the breast itself, meaning that a whole class is named after a characteristic possessed by only half of its members, and among these for just a short period, if at all. In her groundbreaking paper 'Why Mammals Are Called Mammals', historian of science Londa Schiebinger explored the social and cultural forces underpinning Linnaeus' use of names. The question of our relatedness to other species was a pressing one in his time; writing directly into those shifting relations with Nature, he would have felt the need to emphasise our separateness from other animals. In an earlier version of his taxonomy he'd used the Aristotelian term 'quadrupeds', but this had prompted outrage among his peers, who took issue with rational, thinking man being associated with hairy, four-footed beasts. The task, then, was to describe our place among other creatures while highlighting what made us distinct from them. Along with *Mammalia*, Linnaeus introduced a further grouping, *Homo sapiens* – 'man of wisdom' – to describe the species category that humans occupy alone. Consequently, in the language we use to look at and imagine other species, and the one that has framed scientific thought for centuries, a maternal characteristic ties us to other animals, while a purportedly male one sets us apart from them.

 Linnaeus wasn't the only one concerned with lactation at that time. He was a practising physician, working during a period when governments were concerned that Europe's population was declining just as extra labour was needed for military and economic expansion. The French economist Marquis de Mirabeau blamed depopulation on, among other things, the neglect of mothers for their children; breast-feeding was about more than health, then – and the issue of maternal duty had become a matter of the state.

Linnaeus' choice of terminology, and the subtle emphasis it placed on the naturalness of breastfeeding, aligned neatly with these wider changes. Consciously or not, he had called up the ancient association between women and animals and fed it into the political climate of the time. In associating females with mothers, and mothers with beasts, he had invoked an increasingly popular notion of what was natural and normal, reinforcing the image of the simple, unthinking mother, and the belief that her proper role lay in nursing her infant, at home.

—

Our first full day at home and we'd drifted through the morning – everything bathed in fresh light, all sense of to-do lists disappeared. On waking in a tangled knot of blankets, my boyfriend had gone downstairs and come back with fresh juice and a bowl of porridge; now, as I sat in the light cast by the front room window, he made lunch. There was a sense between us, I suppose, that something immeasurable had just happened; something for which we had both been preparing, and in which we each now had a role. We had been imagining this scene for a long time. In fact, we had dwelt in its imagining, studying the picture in detail, and there was a relief, now, in a way, to putting an end to all that, faced as we were with her sudden and breathtaking immediacy.

The baby slept in a Moses basket as I opened the cards and parcels that had arrived while we'd been away. There were soft toys and oversized babygrows. A floppy bunny in a sealed plastic bag, accompanied by a note: *I haven't been touched!* People were worried, suddenly, about contamination. Apparently the post was safe if you quarantined it for a week first, or sprayed it with disinfectant, and at the very least you should wash your hands in soapy water after opening, for a minimum of twenty seconds.

I washed my hands, and washed them again, then sat down in readiness for the next feed. Only an hour had passed since she'd finished the last one. Was a break not permissible by this point? Might I not get some air, get out? But it was only sitting here, being done to. Should be relaxing, should it not? I pulled up my shirt and lifted the baby inexpertly to my breast, where she latched and dozed. In the corners of the ceiling, spiders hung on invisible threads – they'd come in from the fields while we'd been away. Leaning back against the cushions, I watched as flies caught, pulled and wrapped themselves in webs.

A little while later there was a knock at the door, and a midwife entered with a set of weighing scales; brusque, no-nonsense, PPE-ed. A national lockdown was about to be announced, and though we didn't know it yet, this was the last visit we'd be getting from anyone for a while. Standing squarely in our tiny front room, she ordered the baby's nappy off, then reached across and popped her on the scales. —Hm, she said. —Weight's down again.

She lifted the sprawl of spiny limbs and held them to the window. —Yellow, she said, turning to show me. —See?

I looked. There was no mistaking it. The sallow tinge was back, just showing through the pink of the baby's skin. Now the midwife was swift, pragmatic – experienced, I suppose, in driving her point home to parents suffering from a severe lack of sleep. Had we some formula, bottles, a steriliser? Did we know how to use them? Yes, yes, and more or less. Then we were to start supplementing, she said, or we'd end up back in hospital – and I, if I wished to build up my supply, should start expressing after every feed. She slid an anti-bac wipe over the table, another over the scales; pulled off her disposable plastic apron and handed it to me.

And then, she was off.

With the midwife gone, the house was quiet. The street was quiet and the sky was oddly empty. Over the Moses basket, we began bartering. —Could we not wait a day? Increase the feeds, see if that works? —I don't know. —Maybe until tonight? Until the morning?

It felt like failure. Or it felt like a mistake. I stood up and began pacing the room. Three steps and I hit the fireplace; three steps and I was up against the door. Here I stopped and looked back at the baby. She seemed bewilderingly calm.

—Well, how much do you think she's getting? he was asking. —I don't know. —You can't tell? —I can't tell. —Well, doesn't that tell you something's wrong?

Together we gazed at her, hoping for instruction. Something was most definitely wrong, but as to precisely what the problem was and where it lay, not one of us could say.

I turned and limped into the kitchen, pulled the box of formula from the cupboard and stood some moments by the kettle. I didn't want to do it. Perhaps I couldn't. Why could I not? Was it pride, or bloody-mindedness, or the weight of centuries of expectation – some dim sense I had of how this should go, even as it didn't? Instead, absurdly, I climbed the stairs and wired myself to the pump; sat in the bedroom, alone, the baby somewhere below me, my arms emptied, body suddenly weary, determined to get the measure of how much I was able to supply her with.

A draught filtered from the open window. Outside, there were footsteps on the lane – a man, dogs. My breasts were puckered with goose pimples. The nipples had retracted again and the areolas had purpled – the day before they'd been red raw. I switched on the machine and it sputtered into action. *Did you know. Did you know, did you know.* I muttered it now like an incantation, as though I were calling something up, but the images that came to mind were unsettling. Seagulls and sheathbills pilfering milk from the teats of elephant seals; feral cats feeding on the seals' discarded placental tissue, as well as their milk, and their carcasses.

A friend of mine watches YouTube videos of cute babies while she expresses. She says it gets her in the mood. I shut my eyes and thought of animals. I wanted orangutans and chimpanzees, lions and elephants – the kinds of species that, before pregnancy, might have come to mind had I thought of animal mothers. Instead, I found myself thinking next of the leopard tree iguana who abandons her offspring forty-eight hours after hatching, leaving behind a pile of her own faeces – edible and nutritious. And I expressed nothing. A few drops.

Follow the line of thinking implicit in taxonomic language and one arrives at a tidy assumption: breasts equals beastly equals diminished. Linnaeus had parcelled out the animal kingdom, made it logical, and if other mammals were inferior to *Homo sapiens*, further categories were lesser still.

Strange, then, that what I've found again and again as I look to other creatures is not diminishment but complexity.

Venture beyond mammals and the story of infant feeding grows stranger. Some species, of course, don't feed their offspring at all – theirs are born independent. Others gather or prepare food and deliver it to their young. The burying beetle chews up the carrion that forms the walls of its underground nest; the thorn treehopper, a bright-green insect shaped like a single thorn, slices a series of spiral slits in plant stems to expose the nutrient-rich phloem tubes inside. The strawberry poison-dart frog feeds unfertilised eggs to her tadpoles – laced with poisons, they appear to function as a way of passing chemical defences on to the next generation.

In recent decades, milk or milk-like substances have been discovered in far more species than was previously thought. The list, when you see it, is unsettling. There are 'milks' that emerge from the skin, the throat, the vulva; they're emitted, secreted, discharged, released – they're active, and produced not only by mothers, and not only by biological parents.

The tsetse fly, a large biting insect that inhabits much of tropical Africa, produces a nutritive liquid for developing larvae from a specialised gland attached to her uterus. Discus fish emit a milk-like mucus; caecilians, a group of limbless, serpentine amphibians, shed a layer of nutrient-rich skin for their offspring shortly after giving birth. The hermaphroditic nematode *Caenorhabditis elegans* releases a protein from their vulva that destroys their body and ultimately kills them; the jumping spider *Toxeus magnus* emits a fluid from her egg-laying organ to sustain her hatchlings for the first twenty days of life (her daughters will continue to nurse even after reaching sexual maturity).

In some species, both males and females produce a nutritive substance for their young. Flamingo and pigeon chicks are initially fed 'crop milk', a semi-solid substance derived from a sac in the throat. Flamingo milk is red. Parents turn visibly paler through the course of feeding, as their white chicks turn pink. Emperor penguins produce crop milk too, but in this species it is the father alone who feeds the chicks while the mother goes out in search of food.

In the animal kingdom, then, when it comes to feeding, some mothers get help. Even among mammals, there are fathers who play an indirect role in the process by provisioning the lactating female. (Night monkeys beg food from their partners more frequently while lactating; the male supplies the mother, who supplies the suckling infant.) In a number of species, mothers receive feeding support from other females through a phenomenon known as 'allonursing' – observed in over 68 species, this sees individuals overriding the physical costs involved in lactation to feed young that don't belong to them. Giraffes frequently feed each other's offspring; in one study, 86 per cent of calves were observed allonursing (they were more likely to do so when an allomother's own calf was suckling, and tended to adopt feeding positions that made them harder to recognise; the probability of successful allosuckling was also higher if a calf's biological mother was herself an allomother). There are even cases where individuals who don't have offspring of their own will produce milk for another's infant. Among the dwarf mongooses of Tanzania, females who have never become pregnant will lactate to care for offspring in their pack. In a marine science centre in Vancouver, two captive beluga females were observed nursing another's calf over a period of almost three years; the infant was not orphaned, neither allomother was pregnant and neither was nursing a calf of her own. They had apparently spontaneously lactated to assist in the feeding – it's possible, the researchers noted, that such behaviour is also present among beluga populations in the wild.

Far from being the sole responsibility of the static, simple maternal female, infant feeding in the animal kingdom is wide-ranging and fine-tuned, and it can sometimes be shared. There are parents that break rules, unsettle boundaries; animals weirder than I'd thought. Which causes me to wonder what a more extensive, more ecological sense of motherhood might bring. What if we allowed for complex relationships, and for mothers cleverer than we thought? What might it mean for how we relate to other animals, and make sense of ourselves?

I don't remember which of us decided to do it – to make up that first bottle. Perhaps it was him, not me. Perhaps in the end he gently wrested the decision from my hands, and perhaps after all I was relieved about it. What I do remember is the sight of her feeding from the other side of the room; the soft glow of the lamp as he held her, and how thirstily she drank, until the milk was gone.

The next morning, she was pink.

There have always been women who didn't breastfeed, out of either choice or necessity. In Linnaeus' time, more Europeans than ever employed wet nurses, from aristocrats and wealthy merchants to farmers and artisans – a practice he campaigned against, arguing (correctly) that colostrum contained important health-giving properties, and (more dubiously) that the wet nurses of the lower classes ate too much fat, drank too much alcohol and had contagious diseases. According to this logic, the milk of wet nurses was inferior to that of the upper-class mothers, who should, in keeping with the political mood of the time, give up their public roles to breastfeed their offspring themselves.

The other alternative, of course, for those who can't or don't want to breastfeed, is to take the milk of other species – a practice that was in place across the world well before Linnaeus' time, and beyond it. Before techniques were developed for hygienic storage and transportation, the safest way to feed a human infant with animal milk was to do so directly – from the teat. By laying a baby alongside a sow; by holding him to a goat or sheep. In her book *Milk: A Local and Global History*, Deborah Valenze describes how in the eighteenth and early nineteenth centuries, some orphanages kept livestock for precisely this purpose. On visiting an infant hospital in Aix-en-Provence, one doctor reported, 'The cribs are arranged in a large room [. . .] Each goat which comes to feed enters bleating and goes to hunt the infant which has been given it, pushes back the covering with its horns and straddles the crib to give suck to the infant.' Pierre Brouzet, the physician to King Louis the XV, wrote of 'some peasants who have no other nurses but ewes [and their children are] as strong and vigorous as others'. Goats were especially popular, according to Nicholas Day's book *Baby Meets World*, because their nipples were easy to latch on to, their milk was plentiful and they formed strong bonds with children.

My daughter was made healthy by cows' milk. Does that make her more beastly or more a product of the modern world – an era in which milk is a loss leader in supermarkets, and little foil packets of dried-up powder have come to bridge the gap between our infants and the animals that feed them?

The truth was that as I made up the formula, I had to make a conscious effort to consider the cows at all. In those first weeks I was not thinking of animals, of animal stories. I was too intent on getting food into her. Soon I'd learned to make up the mixture one-handed, dipping the plastic spoon into the powder and levelling it out, then tapping it into a bottle and stirring. Soon I was able to time this tapping and stirring so the formula was already prepared when I offered her my breast, the bottle poised and waiting within arm's reach. *If you can't, I can.*

And quite possibly I couldn't. Sitting there, scoop of T-shirt tucked under chin, breast visible to the world, I still struggled to tell if she was getting milk from me or not.

Sometimes, walking into the room, my boyfriend suggested I *do* something while I nursed. A meditation, a breathing exercise, something to send her a message of equanimity. I looked at her so intensely, he said. Too intensely. Perhaps I was putting her off. But I could do none of the things he suggested – it took the whole of my attention just to sit with, and through, the sensation of being suckled (which was, I thought, not unlike being protractedly stung). No flight of imagination, no leap of thought was possible now, and so I studied her. I studied her lips, my breast, the precise angle of her chin and neck, the tender positioning of my arms around her. Were the right parts of me working, I wondered. Was I (God, let me be) natural, motherly, wanting this enough?

It was surprisingly easy, in those early weeks, to feel at sea. It was surprisingly easy to feel alone. And yet, unbeknownst to me, I had just entered into a direct and vital relationship with a host of other organisms. Not just the cows at the other end of the production line, made invisible by their distance; not the spiders hanging from the ceiling. Also creatures inside me, too tiny to see, who would be revealed only when I loosened my sense of my own primacy and softened the lines that separated us.

Until shortly after the turn of the last century, human breast milk was considered by most doctors to be sterile – as though, in breastfeeding, mother and infant exist as a closed loop, untainted by outside influence. We now know that milk possesses its own microbial ecology – a complex mixture of bacteria, yeasts and viruses, all of which contribute to the healthy development of a growing infant. As I'd understood it, the difference between formula and breastfeeding was simple: either I outsourced or took up the task myself. But this was a false dichotomy. There is *always* more than one species at work in feeding our offspring; in breastfeeding, there are many hundreds of them. As mothers, as humans, we are not separate, or set apart. We are always working in collaboration. We are always getting help.

Perhaps this was what I'd missed at the breastfeeding class, and what the neat nurse with her bagful of beads had been so keen to impress upon us. The clever activity of the breast and of the body; the live, constant interaction that goes on between mother, baby and outside world. Scientists don't yet fully comprehend the role of microbial communities within our bodies, but they do know they're essential. In his book *Dark Matter*, colorectal surgeon James Kinross describes how laboratory mice raised in germ-free environments develop strikingly different physiologies from normal, healthy mice. Their growth is stunted, the first part of their colon is enlarged, their small intestine is undersized, their immune system is severely altered. Bacteria appear to have a hand not just in maintaining good health, but in laying the building blocks for our basic physiology. We rely on these manifold, supposedly simple organisms. They make us, they bear us, again and again.

The microbiota of breast milk is surprisingly diverse and contributes to a microbial bloom in young infants – 'friendly' bacteria rapidly colonise a newborn's intestines, and because there is so little competition, these early communities thrive. Alongside aiding healthy development, they help prepare infants for the world, since 'good' breast milk bacteria contain enzymes that enable them to break down solid foods when they start weaning. (A similar mechanism exists in the leopard tree iguana, who leaves her young samples of her own bacteria-rich faeces, and in the panda, whose poo contains a bacteria that helps her cubs metabolise cellulose, the nutritional value in bamboo.)

Manufacturers of infant formula have attempted to recreate the role of bacteria in breast milk with prebiotics and probiotics, but milk is far from the only route of entry for microbial agents. In fact, recent research has detected microbial signals even before the point of birth, in foetal guts, skin and lungs – findings that undermine the traditional assumption of a pristine, germ-free uterine environment.

Whether or not there is active microbial presence *in utero*, a further significant exposure route for microbial organisms occurs during birth, via the vaginal canal – or, if born via C-section, from the skin. Which means that as we held our daughter to feed her, from the bottle or the breast, we passed her our communities – formed from our separate backgrounds, our past experiences. The species living in his chest hair, and the ones in my elbow; the creatures in his collarbone, and in the palm of my hand. They were active, and though they were not entirely benign, they were essential to her growth. They would become part of her, creaturely and teeming – I had no inkling they were there.

———

It was as if, since the arrival of the formula, the baby had discovered a new muscle. Now she whimpered and drank, taking in everything – air, milk, bottle teat – and so hungrily that her little body would contort afterwards with trapped wind, whereafter she'd have to be slung over a knee or shoulder, our fingers learning to seek out the little bubbles of tight air inside her and release them.

And there was an ease to this, to this mode of feeding. It created a greater sense of equity between him and me – of possibility. I imagined skipping a feed, sleeping a little longer; heading out for some hours alone. Still, I persevered with the breastfeeding, even as I came to dread the catch-phrases recited by its proponents.

'Formula is OK, but breast is best,' one friend chimed in via text message, after I mentioned I'd been supplementing with formula. Another, trying to offer encouragement, kept telling me 'it gets easier' – the implicit message being that I should keep at it, when all I wanted was permission to admit defeat, and be forgiven.

'It shouldn't hurt' – this I heard repeated everywhere from the doctor's surgery to breastfeeding manuals to the woman at the other end of the helpline that I called one day from the bedroom floor, baby legs akimbo across my stomach. *It shouldn't hurt.* Meaning if it hurts, you're not doing it properly. Meaning (since I could find no way of nursing except through stinging, singing pain) I was not only hurting, I was also failing, again.

—Are you giving up?! the woman on the helpline asked, in a tone that I think was supposed to be motivational. I was not sure if I was giving up or not. Sometimes I thought that I was. Yet the baby continued to suckle, showing no sign of distress or frustration – only a steadfast and uncomplicated resolve, which she readily applied to whatever teat was offered her. And so I continued to offer myself, perhaps in part because I didn't know how to begin being a mother, and to put my daughter to my breast – to keep doing that, day in, day out, each time the alarm on my phone pinged – seemed as good a way as any of demonstrating this new role to us both. It was an action, like learning to change a nappy or burp her or ease her little legs through the holes of a babygrow; a motherly thing that I could *do*, while the rest of me caught up.

Does it matter what is 'natural' when it comes to feeding? Does all my emphasis on feeding in other animals risk missing the point that in human societies parents should be supported to make informed choices about what works best for them and their baby?

In her book *Matrescence*, Lucy Jones explores the practical, social and psychological barriers that can impede our right to choose. Without access to safe water and sterilising equipment, formula feeding is highly dangerous; without safe and welcoming public spaces and a compassionate work culture, breastfeeding soon becomes impossible. But here I find myself looping back to animals again, because perhaps this is just the kind of perspective that a sustained, closer attention to mothering creatures might bring. Looking beyond the image of the passive, automatic maternal being, we begin noticing her particularities and her contingencies; we grasp a sense of her rights and needs, and what makes it possible for her to thrive.

My animal side, my animal side. A squashed-down, tugged-at, endlessly imagined thing. I see it increasingly as plural and part of us, microscopic and vast – powerful enough to bring about our subjugation, or our freedom.

Consider this. For most mammals nursing is largely instinctive, but for many primates, including humans, it's learned socially. Which means that where we don't grow up around nursing mothers, or where we're isolated from the support of wider communities and friends, women who are trying to breastfeed can run into difficulties.

It's so simple, now I think of it. In pregnancy, sitting in cafes with breastfeeding friends, I felt such an impulse to move closer. I wanted to study the mechanics of it, I wanted to look and look – but I was too polite, I felt that I shouldn't, and so I averted my eyes as they tipped out their breasts, tilted their nipples, rotating their babies to find a latch. In lockdown, I wished badly that I'd been bolder. Now I was making do with videos on YouTube, scrutinising the breast-feeding positions of nameless strangers like I had wanted to my friends'. Some zoos, as it turns out, have taken this same approach with captive primates, screening breastfeeding films to females isolated from their social group. When keepers in Columbus, Ohio, announced a plan to do this for one pregnant gorilla, they were inundated with calls from local mothers offering to give the demonstration themselves. The women came and breastfed their babies along the boundary of the enclosure. They wanted the gorilla not to be alone, I suppose – they wanted her to witness breastfeeding in real time, in the flesh.

But if social barriers to breastfeeding can be easily distinguished, there may be other factors influencing lactation of which we are less aware, and that reveal more complex questions about how we live. In the 2010 UK Infant Feeding Survey, breastfeeding rates at birth were 68 per cent, but by six to eight weeks they'd dropped to less than 50 per cent; at six months, just 1 per cent were breastfeeding exclusively. Among those that had stopped, eight out of ten mothers reported having done so before they wanted to. The same pattern has shown up elsewhere; in one US cohort study, nearly half of all participants said they'd stopped earlier than planned.

It's a curious trend, and one often attributed to the kinds of social factors described above. But many of the mothers in these studies were highly motivated about breastfeeding and did not lack support – instead, they reported giving up because they just didn't seem to be able to produce *enough*.

This perception of a physical, rather than a social or psychological cause, is often dismissed by medical professionals – which is striking, given how frequently as new mothers we're told to trust our instincts. Why trust every instinct except this one? And what if these women, sensing that something is physically awry with their bodies, are not mistaken?

While the dairy industry has spent decades funding extensive studies of lactation in cattle, our understanding of the factors affecting milk supply in humans is surprisingly limited. But it's an area of emerging science, and one that increasingly recognises the role of factors that lie outside the mother's sphere of influence. New areas of research include the field of genetics, and environmental exposures – toxic chemicals, airborne and waterborne pollutants. Or, to put it another way, the maternal body in live relationship with her wider contexts. It's a line of inquiry that, like the findings of microbial traces inside foetal tissues, calls us to consider the ways that mothers are connected and porous, encroaching and encroached upon – radically open to their environments.

But I am running ahead of myself. These are things that in the first flush of motherhood I had not yet discovered. I was still reeling with the shock of her separateness from me, and you have to cleave from one reality, I suppose, before you can begin finding the joins in another.

—

I didn't dust the cobwebs that hung from the ceiling corners, and neither did he. My body was still recovering from the operation, and I found it difficult to stretch – but it was not that really, nor was it that the lack of visitors meant the housework fell to waste, nor that I was too focused on the baby to notice them. I did notice them. A spider nesting over the shower had just laid a clutch of eggs; so had the one above the door. I noticed them and I wanted them there. As I held the baby, and carried her from room to room, I liked the thought that these mothering creatures might also be watching me.

I was, I realised, a little afraid of the baby. I'd expected to feel a sense of recognition – to see in her face some sign of familiarity, some feature that belonged to me – and I did not. She was so . . . *other*. Leaning over the changing mat, I moved my face closer to hers. Eight to ten inches, a little more than a handspan, and the distance at which a newborn is best able to focus. As I gazed, her limbs coiled, her mouth worked, her eyes roved as though without a driver. She gurgled. I grinned.

I was still unsure if I was breastfeeding or not, but slowly, slowly, something else was moving between she and I. At three-hour intervals, we sat together – in warmth and softness; in friction and muscle ache; in daylight and in the dark. These, I now think, were some of my first felt impressions of motherhood. This was, for me, where it started; where I believe we began to know each other, and where we began in a different way to attach.

On the changing mat, the baby kicked. I put her nappy on. It came undone. I tried again. Through the window came sounds of birdsong and a warm breeze. Were there more birds this year, or did it just seem that way? I'd never known a spring like it. The trees seemed alive with nests, with nesting birds, with little pirouetting flights in and out, towards and away and back in.

I was glad the baby was unfurling at this time, becoming aware that she was part of a larger world, that she shared it with these other creatures. But of course she did not yet feel separate from it; that was my mistake. Self, other, here, there – all were continuous to her. There were no edges, no breaks.

IV

FOREVER MILK

Her vomit was a pale yellow. Sometimes, when it dried, it was almost lurid. She was sick constantly, sometimes so soon after feeding that the milk must barely have touched the lining of her stomach before being ejected in a motion that seemed as automatic to her as either yawning or breathing.

These eruptions were normal, according to the health visitor, but they drenched tea towels and muslin squares; they left little flecks and snaking trails along my back, front and sleeves. Soon I gave up trying to stay clean. It wasn't like I was going anywhere anyway. The vomit didn't smell, so far as I could tell, but my milk did, with a stench that seemed to pervade my clothes, our bed, the bath towels – I washed them and I washed myself, but the smell hung around and the milk kept coming, bleeding through nursing bras and T-shirts, pooling in the sheets.

Lying in bed one day, I tried counting the panoply of stains that had blossomed across the fabric: milk, blood, piss, sick . . . it wasn't easy to tell any more what was mine and what was separate. Perhaps, in truth, I had begun melting. Perhaps in splitting open and turning into two like this I had begun slowly degrading; perhaps I truly was dissipating

– into her, into these sheets, into the spring sunshine, the hills.

Dissipating, and being flooded. Milk, vomit, nappies, dirty laundry, someone else's needs . . . if there was a thrill to this influx, there was also a terror. Soon our little kitchen was repurposed as a workstation dedicated to keeping bacteria and viruses at bay. On the surfaces, a series of tea towels for bottles dried, dripping, and waiting to be washed. Beyond them, an electric steriliser – a bulky contraption borrowed from a friend, with an ill-fitting lid that tended to pipe clouds of hot steam part way through the cleaning process. I worried about this. Responsible for making safe the journey from foil packet to baby's mouth, I worried our equipment was not efficient, nor effective enough. Lying awake at night, I would add it to a list of things to do the next morning: *fix or find replacement*. But these were the silent harrowings of the early hours, when if she had not woken, my body would wake me – alert, too wild, strange in its fullness, and I would hold myself still; thoughts ticking inwardly as I listened for the sound of her stirring, whereupon I would fold her to me with relief, and with relief soften into the currents brought on by her feeding – powerful as any emotion, or not separate from them.

Gently, then, I would become aware of sounds outside. Rain on the windowsill, a car door slamming, the day almost begun. Eventually, in the dark, the flooding abated. Or the flooding no longer mattered. Everything flowed into everything else. And with that thought I would tend to follow her lead, and rest gently back to sleep.

FOREVER MILK

―

Some people develop a kind of vertigo as new parents. Carrying their baby up the stairs, or past the sill of an open window, they're drawn by a sudden impulse to throw him, themselves, off. A frightening, dizzying sensation – or perhaps hypervigilance as adaptive response. By envisioning disaster, our attention becomes focused on the source of threat, and the need to steer well clear of it.

On the landing, it is easy. One can step back and the threat is dealt with. More difficult is the realisation that the drop, the fantasy fall, is not the end of it. There are hazards everywhere with a new baby. I understood that I was not to leave teddies in her cot, not to put her down on her tummy, not to fall asleep while feeding – yet she fed at every hour of the night, so of course I fell asleep while feeding.

I had not yet replaced the steriliser. But perhaps I could fix it. The morning was bright and fresh and I was stationed on the sofa, the baby assuming her now familiar position at my breast as my boyfriend left the house for work. A further issue was the limescale crystals coating the steriliser's base. These would have to be removed, but this was also something I could do – in fact, once fulfilled, the task would qualify me for a tick on the invisible test sheet I now carried around in my head. (One day, when these ticks had reached a critical mass, I intended to exchange this test sheet for a dose of maternal confidence.)

As the baby drank and drifted off to sleep I used my free hand to tap words into the search engine. *OK to use limescale remover in bottle steriliser?* Then *safe cleaning, electric steriliser.* I scrolled, tapped and typed again. *Cleaning chemicals, baby, risk.* Here, the search results shifted – I saw that I was no longer scanning household cleaning tips but toxicology reports.

As the baby sighed in her sleep and as nothing tangible in the room moved, I leant against an imaginary wall and fell through it. I had not realised I was pushing, that I was leaning against anything, until the moment it was gone. Here, now, was a new phrase: *endocrine-disrupting chemicals.* A collective term for any substance that interferes with the way a body's hormones work. I swallowed. Endocrine equals hormones equals motherhood. Equals baby. Equals this lizard-like thing, curled up right here in my arms. In one arm, one hand. With the other I typed again, more quickly this time: *risk, contamination, pollution; womb, baby, body; chemical, mother, animal.*

Page after page after page. I read and scrolled as some basic part of my grasp on the world, my sense of my own body, gave way.

Here were mothers invaded and leaking – the stuff of human industry making its way into breast milk, amniotic fluid, bones. The baby stirred. I soothed her. This was a different kind of flooding, a different kind of inundation, swimming through our home and beyond it. I scrolled again, and typed again, remembering the animals whose stories I had whispered to myself and to my daughter; the mothers who had comforted me with their wild complexities, their slow and carefully evolved niches. But there were pesticides in seabirds. Flame retardants in humpback whales. Industrial solvents in penguin eggs, and in seal pups whose bodies are contaminated at birth with chemicals that were banned decades ago.

And then in a moment she was awake again, needing burping and changing, and there was the question of how to eat lunch, and how to wrest a toilet break from the indeterminate period between now and whenever she next fell asleep. I cooed and bobbed one-handed around the kitchen, balancing bottles upside down in the steriliser, forgoing the difficulty of adding butter to toast. Beneath, inside, a feeling like an after-image, turning over, turning the kitchen strange. The box of organic formula, the steriliser in the corner and its tea towel production line – all my attempts at keeping her safe and pure. I wiped a bottle. The steriliser hissed. I had a sensation I get sometimes, that I remember from being very young, of someone laughing at me.

—

The word 'contaminate' comes from the Latin *contaminatus*, meaning 'to defile, to corrupt, to deteriorate by mingling', and originally 'to bring into contact' – to contaminate, then, is to corrupt through touch.

Endocrine-disrupting chemicals (EDCs) were responsible for reports in 2017 of male fish in UK rivers that had begun producing eggs in their testes. Early newspaper headlines traced the cause to contraceptive pills, claiming that hormones from the pills had entered riverways through human waste water – as though in exercising their right to birth control, women were effectively denaturing their environments. *The Telegraph* ran a scrambled headline: *Fish becoming transgender from contraceptive pill chemicals being flushed down household drains*. The story was later picked up by anti-abortion groups with the slogan 'the pill kills'. In fact, the fish were not transgender but intersex (born with a combination of male and female biological sex characteristics), and they'd been exposed to a far greater array of EDCs than could be traced back to the contraceptive pill.

While some EDCs appear naturally (phytoestrogens are present in soy, nuts and oilseeds), others are manufactured, their hormone-disrupting effects a surprise offshoot of other more hotly desired features. Some of them are familiar, their names evoking a curious mixture of familiarity and remoteness (they're substances we know *of*, if not always *about*): DDT, glyphosate, phthalates, polychlorinated biphenyls (PCBs), bisphenol A, parabens, UV filters, triclosan and some 'forever chemicals'. In all, there are just over one hundred synthetic chemicals classed as endocrine disruptors under EU law, and tens of thousands more that have never been tested.

EDCs were first identified in the 1940s when the pesticide DDT was linked to declines in bald eagle populations – among other indicators of widespread contamination, females were laying eggs with shells so thin they cracked.

Since then, effects have been reported in every species brought under investigation. And because hormones are the great drivers and regulators of fertility, pregnancy, birth and lactation, these processes are particularly vulnerable to disruption. Fish are especially sensitive, since embryos lack the protection of a hard eggshell, meaning they can be directly exposed during fragile developmental stages – but ultimately the influence of EDCs extends far beyond aquatic ecosystems, and across all taxonomic groups.

Curious, then, that in all the pregnancy and breastfeeding manuals I read, I saw no mention of them. Nor had I come across them in the stories I'd found of parenting in other species, meaning there were things, as I readied myself to become a mother, that I specifically did not learn. I did not learn about the beluga whales pulled from the St Lawrence Estuary between 1983 and 1990, their carcasses so contaminated with industrial pollutants that milk production had been compromised in eight out of seventeen mature females. I did not learn about the rodents in sparkling clean laboratories whose mammary glands failed to develop fully following high exposures to per- and polyfluorinated alkyl substances (PFAS), those forever chemicals – a large group that includes a number of proven endocrine disruptors. I did not learn about the human studies that found that high exposure to forever chemicals was significantly related to early undesired weaning, or not initiating breastfeeding at all; or that they might be present in the make-up I applied to my skin, and the waterproof coating on my raincoat, and the stain-resistant fabric I paid extra for when, feeling very grown-up and proud of myself, I purchased a new sofa for our home – a sofa that I imagined would weather a growing child. The connection – mother, baby, contaminated world – was not often presented, or if it was, the problem was limited; a matter of personal responsibility. Eating less seafood, for example. Avoiding cigarette smoke. The problem was practical; the problem could be written on a list and pinned to the fridge. It was about steering clear of certain foods, certain places. It was not everywhere. It was not all-pervasive. Yet even still, even then, it had bothered me. The thought that the maternal body was not *necessarily* protective; that it was not *wholly* sealed.

When my boyfriend returned home from work that day, the day I'd planned to fix the steriliser and did not (a cross on the invisible test sheet), I was . . . woolly. I had little to say. I repeated the fact of my tiredness, repeated the chores that were still to do. Meanwhile, lower down, I wondered what to make of a world where edges are disregarded in such a fundamental sense. Where we are so cluttered. Where we are so open.

—You OK? he asked. I sat up, readying my reply, and he glanced at me warily. His face was grey with tiredness. It said: *You have been at home all day. You have been at home, with our daughter.* My mouth snapped shut. I straightened my shoulders, swallowed the story of the beluga whales. Was I OK? The answer needed to be a yes because there was no space any more for not being. —Yes, I said.

The feeling was like this: glimpsing a landscape, seeing none of the detail. I had no choice but to move towards it. The baby refused to nap anywhere except attached to me. Therefore there were whole hours of the day during which we sat, stacked together on the sofa; hours when it was possible to swipe, tap and scroll through articles unimpeded. Holding infants like this, chest to chest, is supposed to help them regulate their heart and breathing rate; it is supposed to foster a sense of security. Both these terms – security, regulation – seemed increasingly cloudy to me. Her mouth worked reflexively in her sleep as spiders shifted in the ceiling corners, threading wall to wall to wall.

It is not that edges are *disregarded*. It is that our bodies do not exist as discrete entities – a realisation that can come as a shock because it changes the picture we drew for ourselves: *Homo sapiens*, set apart. But this is how it is to enter the world of chemical pollution; you must learn a new set of rules. Stop seeing the world in terms of distinct objects and think instead of movement, of flow. If I sensed this early on, I have learned its detail much later, now that my daughter is increasingly somewhere beyond me, out of reach sometimes, and separate enough that I can begin putting my mind to other things. 'Rules' makes it sound like there's an overseer somewhere, organising everything, which there is not, so instead I'll call them 'principles'.

The first is porousness.

*

Not only does the body leak, it is constantly being invaded. It contains things originating from outside of itself, synthetic as well as animal – an insight that deals a final blow to the idea that in motherhood we inhabit a realm of pure, untainted Nature.

In her seminal work *Having Faith: An Ecologist's Guide to Motherhood*, American biologist and author Sandra Steingraber charts how the notion of purity in relation to motherhood came to bias our understanding of basic biology, so that for generations the placenta was conceived as an unbreachable barrier, impervious to outside influence. Such a notion would have held little traction among the ancients (Aristotle believed that blood was sent directly through the placenta and into the foetal umbilical cord), or during the Middle Ages (in the twelfth century, Thomas of Aquinas made the now chillingly prophetic claim that whatever enters the mother's body also passes through the placenta). But by the Victorian era, an age that revered pregnancy and motherhood, the placental barrier concept had taken root – indeed, it was so tenacious as to remain in place even as evidence emerged to the contrary (though environmental pollutants were understood to cause birth defects in animals, such slippery passages of entry were not considered applicable to the human body).

Another century on and it was generally recognised that maternal malnutrition, X-rays and some drugs could harm unborn foetuses, but even as late as the mid-twentieth century, medical students continued to be taught that the placenta was impervious to toxic substances. Such a view was no longer tenable after the 1950s and 60s, when a drug initially introduced as a tranquilliser was prescribed as a treatment for morning sickness despite never having been tested on pregnant women. Thalidomide flowed straight through the placenta and into the foetal environment, with devastating consequences. Search the newspapers of that time and you'll find pictures of children missing hands or feet; infants born with stunted limbs. At the time they were referred to as 'flipper babies', a name that itself enacts a disturbance and a violence – the infants deemed at once more animal and less natural.

The point of contact between a mother and her foetus is the placenta's delicate river-like structure and its umbilical pipeline – the mark of our deep ancestors' move to viviparity, which offered unprecedented protection for developing offspring as a trade-off for the extra time and energy involved in carrying a foetus to term. In fact, it is not a 'barrier' at all, but a live filter capable of choosing some chemicals over others – a fine, dizzyingly intelligent capacity, until you realise that the many thousands of synthetic chemicals circulating our environments today have appeared so quickly that our bodies have not yet evolved to distinguish what is toxic from what is safe. Subsection of the first principle, then: the placenta isn't just porous but actively selecting. As it sorts on the basis of molecular weight, electrical charge and lipid solubility, small and neutrally charged lipophilic molecules are actively shunted into the foetal environment.

Like some pesticides. Like heavy metals and flame retardants. Like a whole array of other endocrine disruptors, including parabens, triclosan, bisphenol A (BPA) and its common replacements (BPB and BPS), fungicides, herbicides, PCBs, polycyclic aromatic hydrocarbons (PAHs), polybrominated diphenyl ethers, dioxins and furans and some forever chemicals – along with a number of previously unidentified chemicals that have become known to researchers only through the discovery of their hormone-disrupting effects.

Our time *in utero* is the most vulnerable period of life. It is also the one that we know least of all about. As with their animal counterparts, human mothers were overlooked by researchers for decades; even now, gestation is a notoriously difficult field for investigation, shrouded as it is in the womb's darkness, and bound in ethical red tape. Yet foetuses are more biologically sensitive than any other human group. During development, they pass through 'windows of vulnerability' – critical periods when a particular set of genes is being activated and cell reproduction rates are high, where EDCs can trigger changes along immature and highly malleable developmental pathways, making the precise timing of exposure sometimes even more important than the dose.

Some EDCs are more toxic than others, but a growing body of evidence now associates prenatal exposures with effects that manifest later in life: a predisposition to some illnesses; a reduced immune response; a reduced sperm count, early-onset puberty, male genital abnormalities, accelerated menopause. The list goes on, and it is contested, but it is not unfamiliar. As with the new mothers reporting early cessation of breastfeeding, these are trends I've heard about – subtle shifts becoming visible at the level of a population.

The picture is different in instances of unusually high exposures, such as downriver of the DuPont plant in West Virginia, USA. Here, beginning in 1951 and over the course of the next fifty years, PFOA (a forever chemical now classified as an EDC and human carcinogen) was used in the manufacture of Teflon, before being pumped into the Ohio River or dumped into open, unlined pits. There was little subtlety about the effects of contamination in this community, though DuPont fought for decades to conceal them anyway. Fish were found floating in rivers; deer lay dead in the road; cattle were born with malformations and died; human workers at the plant developed cancers, tumours, leukaemia, suppressed immune systems. PFOA has since been phased out, but its spread is so extensive that every baby born in the Global North today carries its trace in his or her blood. If we are more animal than we have allowed ourselves to believe, we are also more chemical – just as the species with whom we share this world, who have for so long supplied the 'natural' yardstick by which we measure ourselves, now bear within their bodies the physical marks of human industry.

—

Sometimes as she lay in her father's arms I climbed on a chair to inspect the spider nests in the ceiling corners, their eggs morphing gently from translucent little nothing balls to forms of substance and complexity. Inside each egg one could see, or sense, but see, I thought, the shapes of bodies, coiled legs. What would happen when they hatched, I wondered – were we to be overrun with spooling spiderlings? What was the difference between a family and an infestation? Sometimes I pictured the walls alive and crawling with many-legged creatures, and sometimes I was not averse to it.

If on the ground we were battling against chaos, these arachnid mothers appeared to live above it. Envying them this ability and wanting to protect it, I'd continued to leave them be, and as though by way of a reply their webs had proliferated, slung like stretched ropes, knotting to everything and nothing at all.

I desired such work. I recognised it. Sitting on the sofa as she fed or slept, almost reflexively I'd begun making lists on my phone – bookmarking pages and research papers, forming little strings of links that would at some future point connect me with the knowledge I now compulsively sought. This was different from collecting stories of animal mothers. It was more determined, more purposeful. With that first glimpse of the world of EDCs, some impulse had moved, quick as a heartbeat – *how am I affected? How will it affect her, how will it affect me?* – and I intended to buck it. Were I to limit my learning to a concern for myself and my own, I would no more grasp the world of chemical leakage than a spider could form a web from a single point on the wall. I must look farther, then. I must think of more things.

So it was that as my daughter's angular limbs disappeared inside rounds of flesh, and as my own body whittled and grew ravenous, I scanned studies of mosquito larvae feeding on contaminated microplastics; reproductive failures in harbour porpoises; foetal deaths among minke whales. *Male peregrine falcons in Spain producing female egg-yolk proteins. Arctic northern fulmars contaminating unborn chicks with phthalates.* I had heard of phthalates. They were used extensively in children's toys until 1999, when a number were banned on safety grounds.

The spider webs spread, a little farther each week, and the lists grew – and then came the day when the baby would not stop vomiting, nor crying for more, her tiny stomach ejecting the milk again and again, seeming so incapable of holding it down that I called the doctor, frightened, and was summoned down to the surgery without delay.

Scrambled messages to my boyfriend, who was out at work: *They said be ready to go to A&E. Yes will let you know. Yes leaving now.* And then a rush of relief, despite it all, as we left the house – the hills green and bright with summer sun.

The doctor was waiting for us at the entrance, dressed in a plastic visor and a flimsy apron, poised in the open door. I padded after him along a soft yellow corridor, past stock art prints and tubs of hand sanitiser. I wanted to drink it all up: the hum of the radiators, the feel of the carpet, the faint smell of antiseptic. I wanted to breathe it all in. —I'm sorry, I said to his aproned back, though I was unsure what I was apologising for. Perhaps for the failure of my milk;

perhaps for our intrusion; perhaps for feeling so glad to be there.

The health visitor had come too, with a set of weighing scales and a flush of floral perfume. Adjusting our masks, we squeezed into the small consulting room and I placed the baby on the bed. She kicked and grinned. She no longer seemed particularly sick.

Indeed, she was not. The doctor been concerned that the symptoms were down to a condition caused by a narrowing of the digestive tract – the body effectively creates its own barrier to ingestion. But now he checked the baby's tummy as she blew raspberries at him, and pronounced her fine. There was no sign of blockage, and her weight was up. The room calmed, the doctor seemed relieved. He sent us on our way.

Back outside I sat a moment in the car, door slightly ajar, sunlight dancing through the trees. The baby fidgeted in my arms and, for want of something to soothe her, I put her to my breast – whereupon she fed, simply and cleanly, and fell asleep.

A sparrow hopped soundlessly across the tarmac. Not noticing us or not perturbed by our presence, it began picking at something on the road. I watched it blankly, dimly aware that I should be texting my boyfriend. I didn't want to. The thought surfaced and disappeared again. I didn't want to do anything but sit here a while longer, no demands being placed, nothing to do but gaze into the middle distance. The bird hopped closer. Now it seemed to be considering me, or considering the space between us.

A moment passed, and then another. I was so lost inside the stare of this little bird, I ceased to register that it was in a road, and I did not register the car's approach. Even if I had, I would have trusted the instinct of this small, delicate creature for flight. But it did not fly, or not quickly enough. Instead, as the car swung around the bend, the bird flew up, but not away, and bounced. It actually bounced off the windscreen, a reflection of bright clouds and sunshine obscuring the driver inside. I cried out. The baby slept. The car sped on, carrying with it the dusty imprint of a bird flung against its glass.

The second principle is persistence.

All substances degrade eventually, but some do so more quickly than others, and some have been developed specifically to resist breaking down. PFAS, or forever chemicals, are among the most persistent organic compounds on the planet – a group of several thousand substances whose stubborn properties have been used to weatherproof clothing, smother flames in firefighting foam, withstand heat and grime on non-stick frying pans, and waterproof recyclable coffee cups, mascara and nappies. They can endure for decades, meaning that when they enter our environments, through disposal or degradation or as by-products of other processes, they hang around. They're also lipophobic, meaning they repel water and fat, but bind to proteins, such as those in breast milk.

*

Fast-forward two and a half years and I am sitting on my daughter's bed while she plays with my parents outside. Since I can no longer afford the monthly rent at the work studios in town, the back bedroom, now her bedroom, has also become my temporary office. The front bedroom is where I sleep, and where she sleeps too sometimes in the middle of the night, coming through to lay herself spread-eagle across the mattress. I wake up with her foot in my side, her hair in my face. We are rearranging ourselves, she and I. We're taking up more space, adjusting to a new set-up, the house pared, her father a distance away, the three of us spread.

Beside me, on top of the duvet, is a file stuffed with handwritten notes. Finally, I am working my way through the lists on my phone, the bookmarked pages, knotting new knowledge to new knowledge. I tell myself that I am getting back to work, that this is all part of providing for her – yet in truth I have begun replacing real work (teaching work, mentoring work, salaried, *proper* work) with stories of chemical contamination. With penguin eggs and porpoises. Last week I wrote to my boss at the university to ask if I could reduce my teaching load next term – a wild, perhaps irresponsible move. What is it about this act of collecting, of collating, that has become associated in my mind with provision – with security, even?

Lately, I have begun speaking with a toxicologist in Holland; with epidemiologists in Denmark and Scotland. I told them I'm researching for a book. This, I realise, is how a writer explains her strange compulsions, her unwise choices.

It is how she makes them acceptable. Since speaking with these experts, on hushed Zoom calls while my daughter naps, I've set up email alerts for colleagues they mentioned off-hand. Now a little ping arrives on my phone as new reports are published – as I'm reaching for pen lids under the sofa, or spooning beans over toast. Today, as I was getting her dressed, a study showing an inverse relationship between forever-chemical exposure *in utero* and reproductive function in young men. My time is limited, so I have become pragmatic – I click, scroll, clock the key facts then put the phone down again, returning to the business of zipping and buttoning, teasing toes into socks. But no, that is not quite right. Reading these studies alone, after she is in bed at night, invariably I begin to cry. Always, this surprises me. It is the animal stories that do it the most. The thought of all those creatures, those wild creatures, polluted with the stuff of us.

In the garden, I can hear her talking. She's deep in conversation with my dad as they investigate the soil under the plum tree. She has found something unexpected and is forming a theory about it – I can tell from the way she stands, head slightly inclined, intent, vocal (like me/not like me). —Kangaroos have blue poo, she told me this morning, with aplomb. —Do you know what granny milk tastes like? she said as we ate breakfast. —I think it tastes like mummy milk. Always, I am amazed by her eagerness to enter into the world, to form sense from it. How it seems to pull her towards it, and how she answers its pull, not questioning her right to be included, to know.

Inside, I have been thinking at a molecular level. One thing does not simply flow into another, I remember now. A particle is either attracted or repelled; it binds or is rejected; it is drawn into a relationship or it is not.

Having taken up residence in living bodies, some persistent chemicals remain there for decades, and with repeated exposures they accumulate. Meeting with the toxicologist last week, I asked what happened to these substances as they entered the foetal environment. In that fine and watery world, what are they drawn to – where do they go?

She told me about a study that found forever chemicals in embryos and foetuses at every stage of pregnancy – in lung tissues, in livers. And for other persistent endocrine disruptors? For heavy metals and flame retardants, pesticides and PCBs? Unlike forever chemicals, most other persistent pollutants are lipophilic, meaning they build up in our fat stores. And during pregnancy and breastfeeding, of course, a mother's fat stores are mobilised – plunged into the service of the growing foetus, the breastfed infant. So what happens, I asked, if they make it through the placental barrier? Her answer was that we don't precisely know. But we can make good guesses. —Like for example, she said —the fattiest tissue in the developing foetus is the brain.

She and her partner have been trying for a baby. I couldn't help it, I asked her: how did she hold the two things in her mind? Motherhood, pollution. Had she found a way of reconciling them? Her answer was only and immediately that she cannot. The two have to remain separate; when she goes out to work, she leaves the mothering part of herself behind.

In asking the question, I realised that I had been looking for reassurance: *it's not as bad as you think*. But it is that bad. It is so bad that a person must build imaginary walls around themselves in order to live.

Daily exposure to a complex mix of EDCs, including persistent chemicals, is inevitable. The majority of humans now carry them in their blood; the majority of mothers in their breast milk, cord blood and amniotic fluid. Since, to my frequent surprise, I am still breastfeeding, I have been trying to find a lab that will test my milk. Initially I tell them I want to test for forever chemicals, but the more experts I ask, the more pollutants are proposed. One researcher in Norway suggests I try testing for chlorothalonil, which he says is more likely to show in breast milk. I've never heard of chlorothalonil. Looking it up, I learn it's a fungicide, though it's also been used as a wood protectant, and to treat mould and mildew. Another researcher tells me I'd be better off testing for legacy contaminants such as DDT and PCBs. Another says that I am too late, too late – that having breastfed for over two years, tests would be unlikely to show anything at all. Whatever chemicals were present in my milk will already have been transferred to my daughter.

Both DDT and PCBs belong to the 'dirty dozen', a group of highly toxic persistent chemicals that were either banned or heavily restricted at the Stockholm Convention in 2001. This milestone international treaty came into effect in 2004, and in the years since, more chemicals have been added to the list, including PAHs, some flame retardants and a number of forever chemicals.

PCBs are some of the most widely distributed chemicals in the group. Colourless and viscous, invulnerable to electrical conduction or burning, their production began in the 1930s when they were initially used in fire extinguishers and liquid insulators; later, they were poured into the ballasts of fluorescent lights and added to hydraulic and microscope immersion oil, to ink and paints and carbonless copy paper.

PCBs can be held by the human body for between twenty-five and seventy-five years. Their toxic status was first identified in the 1950s and 60s following a number of devastating public health scandals, such as the Kanemi Yusho incident of 1968, where PCBs contaminated a stock of Japanese rice oil. Here, the leakiness of the placental barrier was unmistakable. Of the eleven babies whose mothers had consumed the oil in pregnancy, two were stillborn. Among the nine who survived, skin diseases, respiratory problems, severe headaches and muscle weakness were common. Follow-up studies showed impaired IQs – we now know that alongside their endocrine-disrupting properties, PCBs can harm the development of the brain.

Over 152 countries signed up to the Stockholm Convention, and in the years since it came into force, PCBs have shown an overall decline – but they are still everywhere. North Korea still manufactures PCBs; in several African countries, cooking oil in local markets contains pure PCB formulations, and some Nigerians treat their skin with PCB transformer oils to make it softer and lighter. They're in legacy products, in landfill sites, in river sediment, and they're making their way into our seas. As a result, they continue to have a severe impact on the environment and wildlife – and nowhere is this truer than in the context of large marine mammals. Here, then, is the third principle: biomagnification.

There are terms, when one spends time reading around this topic, that become familiar. *Apex predators*, *sentinels*, *canaries in the mine* – words used to describe the kinds of creatures that in the first flush of babyhood came to fill up our home, emblazoned as they were across babygrows and hats, bottles and bedclothes, their manufacturers seemingly incapable of producing anything without referencing the living world. *Polar bears, orcas, humpback whales.* I dressed her in these creatures; I wiped and scrubbed them clean. But they also populated the reports I scanned as I compiled my lists, so that as I folded and tidied, I was caught sometimes by the sense that those distant creatures were also here – they were also close.

Remember the mosquito larva, mindlessly ingesting whatever drifted into its path? As that larva is eaten by a fish, any chemicals contained in its body are shifted up the food chain. Bigger animals need more energy, so the higher the predator in the chain, the more food it consumes, and the greater the chemical concentration; this is the logic of biomagnification. In the case of persistent chemicals, such a process can occur over decades, leaving some predators – such as long-lived, blubber-rich marine mammals – with devastatingly high loads.

Remember the beluga whales stranded on the St Lawrence Estuary, their carcasses splayed for investigation? Cancers were found in 27 per cent of the creatures necropsied, despite the fact that whales are not usually afflicted by the disease (among another beluga population, living thousands of kilometres away in the Canadian Arctic, no cancer has been recorded at all). The St Lawrence River draws its water from a large area, including land occupied by the many industries lining the Great Lakes. Pollutants dumped in the St Lawrence over decades have settled on the river's bottom and been absorbed by algae and small invertebrates – meaning that as a beluga noses the riverbed, stirring up sediment, she is exposed to pollutants both in the act of seeking food and when ingesting it. As a result, the chemical load of some whales in this region is so high that they are classified under Canadian law as toxic waste.

If one effect of industrial pollution among the St Lawrence beluga population is the high incidence of cancers, a second is their lowered reproductive rate. A third is that even when a female is able to successfully reproduce, she offloads her chemical burden onto her calf, both *in utero* and via her milk; meaning that, before he has even begun to feed himself, his body already holds the pollutants at large in his environment, and those that drifted through decades ago – a strange and toxic kind of haunting. In these circumstances, the firstborn is first in line, and thereby receives the highest load, along with the highest risk of early death.

Again and again, clicking through these studies, I have come across creatures I recognise, and stories of fine-tuned maternal bodies gone awry. Hooded seals produce the richest milk of all mammals; their pups nurse for just four days, during which time they must build up enough blubber to survive alone on drifting sea ice. Imagine what happens, then, when that milk is laced with forever chemicals, which have been associated with low birth weights. Everything shifts. Connections are lost, links fall away and that carefully evolved relationship between mother and offspring is thrown out of balance.

Such an out-of-kilter existence is why a population of orcas living off the coast of Scotland has not reproduced for the last twenty-five years, and why they'll likely go extinct within our lifetime. It is how the veterinary researcher and pathologist who led a necropsy on a thirty-year-old female from the group came to describe the orca's tiny, fibrous ovaries as 'scar-like' – a word that has echoed in my mind, describing as it does not a wound, but something *like* a wound;

not touch, exactly, but a more sinister and insidious kind of contact. She was named Lulu. Her ovaries had never once released an egg. Scar-*like*, wound-*like*. Further testing revealed a level of PCBs in her body at a hundred times the 'safe' limit.

Steingraber describes a moment of insight when as a student she was encouraged by a professor to look more closely at the food chain pyramid posters that had been displayed in classroom laboratories all her life. Phytoplankton, zooplankton, fish, seal, human; a neat diagram showing biomagnification's flow, and one that she had accepted without issue until the day she was urged to question it. Who was at the top? A man. No, who was *really* at the top? Not a man, but a breastfed infant. 'The hard fact of biomagnification,' writes Steingraber, 'means that breastfed infants have greater dietary exposures to toxic chemicals than their parents.'

―

Her cloth nappies were printed with ocean waves. The design went on and on in vivid technicolour, the waves forever cresting – a flood that never ceased. I watched them spin, soap-sudded, in the washing machine; I left them to swing, aimless, in the breeze. I liked them, though their numbers overwhelmed me – the nappy bin was never empty, or there was never enough pause between emptying it and it having returned to a state of almost full. What I liked best was to put her in them, since the moments in which she lay – naked, wide-eyed – on the changing mat were usually the most alert of her day.

Curiously, of that period, these are some of the clearest memories in my mind – times when a few degrees of separation lay between us, and when I could become aware in the most basic way of her presence. Waking at night, listening out for her breath; stepping into the garden, alone, the world brimming, only to feel a tug back inside, back towards her. And then those moments on the changing mat – the sight of her unencumbered, kicking.

There are vaguer memories, too. Blurriness, weariness, a permanent juggling between the now and the just ahead. At almost three months she rejected formula, refusing to feed from the bottle in a sudden and surprisingly defiant show of will. My breasts lost all recourse to assistance – and there was a new keenness about her as she persuaded my body into more, and more again. It was true what I'd read – she was taking from my thighs, my bum, my waist. As she ballooned, I waifed. I ate and I ate. My internal body temperature rose; I had never felt so productive, nor so ravenous. Passing her to me in the bath one day, my boyfriend's hand grazed the side of my breast, whereupon he sprang back in surprise. —Jesus! he gasped. —It's *hot!*

And it *was* hot, or warmer than any other part of my body, as was the milk, which sprang from my breast like a garden sprinkler if ever the baby lost her latch, her attention caught by something else.

Now I kept a muslin square slung over my shoulder not

for wiping her vomit, but for mopping up my own leakages – which, alongside whatever persistent pollutants I'd accrued through the course of my own life, likely contained chemicals absorbed into my mother's body long before I was born. It was freighted, this milk, and it made me at once porous and dangerous – a far more complex being than the benign entity I had wished myself to be.

This capacity of milk to become a vessel for other substances was recognised back in the nineteenth century, when nanny goats fed with mercury were used to treat babies born with congenital syphilis (while the mortality rates of the suckling infants improved, the goats tended to die of mercury poisoning); but it wasn't until the middle of the last century that evidence emerged of pollutants at large in the environment making their way into human breast milk.

The link was made via DDT, the highly toxic insecticide that Rachel Carson would go on to write about in *Silent Spring*. In 1949, a doctor working in Connecticut noticed that a woman with acute DDT poisoning began showing signs of recovery following the birth of her child. The symptoms, which included vomiting, abdominal pain, hyperirritability and muscle weakness, started to dissipate after breastfeeding was established. On analysing her milk, the doctor discovered it was highly contaminated with DDT – her body had begun shunting its chemical burden onto her newborn infant. Two years later, researchers tested the breast milk of thirty-two women at a Washington, DC hospital. Those taking part in the study had not worked with insecticides or reported high exposures to DDT, yet the chemical was present in all but three of the milk samples tested. This was the first evidence that even low-level exposures, likely through diet, could be offloaded by the maternal body via milk.

Over twenty years later, in 1966, Swedish researchers testing for DDT and other pesticides in wildlife came across a series of 'unknown compounds' they later identified as PCBs. Further investigation revealed PCBs in fish and seabirds, prompting one of the researchers to screen his own wife and children. They too contained the chemicals, including his five-month-old daughter, who had been fed solely with breast milk.

Despite their production and use having been heavily restricted in many parts of the world since 2004, PCBs and DDT continue to be found in breast milk, and they are joined by an increasing number of other pollutants, including forever chemicals – though the precise concentrations and mixtures vary widely depending on where a woman lives, her age, lifestyle and socio-economic status. In fact, breast milk offers such a clear marker of toxic substances in the body that it has often been used as a biomonitoring matrix to assess population exposure levels.

It's too late to tell what chemicals I passed to my daughter via my breast milk, and if there were forever chemicals among them. Three are currently listed under the Stockholm Convention (PFHxS, PFOA and PFOS), but thousands more have been registered for manufacture despite very little testing for toxic effects. Among those that have been studied in depth, all have raised profound health and environmental concerns.

In the press, much has been made of the fact that forever chemicals are present in some cosmetics and skincare products – that we are, mostly unknowingly, rubbing them into our bodies. But as a route to exposure, cosmetics may play a relatively minor role; PFAS are also entering our bodies through food and drinking water, and via household dust.

There is something particularly unsettling for me about a pollutant that collects in dust. House dust belongs to interiors – it's in corners and under sofas, down the back of beds. It encroaches, it builds up in our homes. But forever chemicals are outside, too. They're ubiquitous in rainwater at concentrations above the safe drinking limit; they're in river systems and soils, and therefore in our food. They are, quite literally, everywhere. Yet it is so difficult to get a hold on what 'everywhere' means; to grasp the intrusiveness of that word, and its extensiveness. Everywhere is in cushions and carpets and in the cracks between floorboards; it's in our blood, and in the blood of other creatures, and in the furthest places we've travelled to, and the ones we'll never see. To try and understand this is to get a little closer to the flood, I think. It is to let go of the idea that our bodies are protected against incursion; to grasp that we are as contaminated as our environments, and as vulnerable.

*

So what to do as parents – how to keep our infants, our unborn children, safe? The thing about persistent chemicals is that no amount of healthy eating in pregnancy can alleviate what has already accumulated over decades. The longer a mother lives before having children, the more chemicals have built up in her system; the longer she breastfeeds before weaning, the more offloading will occur and the higher the infant's overall exposure. If over the course of her lifetime her diet has been rich in seafood, this exposure is likely to be higher again. Older siblings will duly receive a greater chemical load than younger siblings; a mother's body mass may also play a role. So, maintaining a vegetarian diet, having lots of children early in life and weaning early can all reduce maternal offloading through milk – but, for many, such choices are not practically feasible, or desired.

I wondered sometimes if perhaps I shouldn't be breastfeeding at all. Infant formula tends to contain lower levels of persistent chemicals than human breast milk, since cows are lower down the food chain, meaning less biomagnification has occurred. But of course this isn't a simple equation – and the story doesn't end there. The sterilising process, highly effective at removing bacteria and other germs, causes plastic bottles to shed millions of microplastic particles and trillions of smaller nanoplastic particles, meaning that most bottle-fed infants are consuming somewhere in the region of a million or millions each day (compare this to an estimate by the World Health Organization that adults consume on average around 300–600 microplastic particles per day).

While many of these particles will be excreted, we don't yet fully understand how many are being absorbed into the body, and to what effect.

In the Global North we're accustomed to assuming a certain protective factor when it comes to environmental crises, but in the context of environmental pollutants the tables can be unexpectedly turned. The UK, Europe and the USA have some of the highest rates of formula feeding in the world, meaning that infants born in these regions are likely to be receiving the highest levels of microplastics exposure. In the context of forever chemicals, too, higher-income groups tend to receive higher exposures, since they are the ones buying lots of new things: expensive ski wear and outdoor gear, new furniture. One researcher I spoke to told me that if I wanted to limit my exposure, one way of doing so was to buy furniture second-hand, since their stain- and water-resistant coatings are likely to have worn away. Another told me that weathered, degraded fabric on second-hand sofas might be more likely to release toxic microfibres; a third that older furniture may contain chemicals that are now banned (in the UK, some furniture can no longer be landfilled, reused or recycled since it contains pollutants restricted under the Stockholm Convention). Elsewhere, I learned that it's possible to reduce exposure to forever chemicals inside the home by hoovering regularly – though evidence on this point is contested too, and some argue that frequently unsettling dust can actually increase overall exposure. It seems there is no getting away from them.

*

Just as foetuses are particularly vulnerable to EDCs because gestation is a time of intense development, so infants are more sensitive to any toxic effects than adults because the first years of life are characterised by rapid growth and change. In the first three months after birth, brain volume increases by over 60 per cent; the brain, lungs and immune system all continue to develop throughout childhood. But start unpicking the potential adverse health effects of breast milk contamination and the issue quickly becomes tangled, since breast milk is also brimming with health *additive* effects. EDCs interfere with the development of healthy immune systems, but breast milk helps build healthy infant immune systems. PAHs, a group of persistent organic pollutant compounds and endocrine disruptors produced by burning coal, oil, gas, wood, rubbish and tobacco, can put an infant at greater risk of developing lung and respiratory problems, but breast milk has a protective effect against these very conditions, and so provides a kind of advance shield against the kinds of environmental exposures a child born to a mother with PAHs in her milk is likely to encounter.

The World Health Organization has been reluctant to broach the subject of breast milk contamination, since for people living without access to safe drinking water, breastfeeding is essential. In the scientific literature, the consensus is that for those who want and are able to breastfeed, the benefits of human milk outweigh the possible risks of contamination in all except the most extreme cases. This means that, at present, unless you've received unusually high exposures, the EDCs in your breast milk are unlikely to do much more than nudge the risk factors affecting your child. Perhaps they'll be slightly more susceptible to infection; perhaps less abundant in sperm. The Danish epidemiologist I spoke to chose to breastfeed her child, despite never having tested her own milk. For her, it was an emotional decision – and one that she didn't share with her supervisor at the time. The adjustment to motherhood hadn't been easy for her; breastfeeding was hard won, and about more than the milk.

I relate strongly to this. Motherhood is not an entirely rational undertaking; we can't always explain why we do what we do. I don't know if it was nature or culture that made me so eager to breastfeed my daughter – if it was an inclination I'd absorbed from the people around me, or some deeper, more primitive instinct. What I do know is that sitting here, attempting to parcel out careful calculations of risk, is a madness. I'm not equipped to make these calculations, and neither (as they themselves attest) are the experts I've spoken with. I shouldn't be here, I shouldn't be writing this. No mother should be put in the position of having to decide whether her milk is unpolluted *enough*; it shouldn't be polluted at all.

But it is. And so I have also been speaking with a veterinary epidemiologist in Edinburgh. He and his partner made a different decision some years ago, opting not to become parents; not to bring a child into this overheating, increasingly toxic world. —It was difficult, he tells me, when we speak over Zoom at the end of his working day. —There were long conversations, over many years. But, on balance, I have to say it was the right decision. It was the right decision for me.

He is choosing his words carefully, I notice. I wonder if he is afraid of what he might do to me, of what I might do.

He heads up the marine animal stranding scheme in Scotland – it was he who led the necropsy on Lulu, the orca stranded off the Scottish coast; he who used the word 'scar-like'. He estimates that 80 per cent of the world's PCBs are still stockpiled on land; over the next few decades they'll leach into marine ecosystems. This is why those whale mothers, the great humpbacks, the orcas and belugas, are being called canaries in the mine – because the crisis is still to come; the flood is still heading this way. Under the fluorescent bulbs of his university office, this researcher looks very tired.

Friends have been wary when I've told them I want to write about these things. Perhaps, they've suggested, the reason EDCs are not mentioned in pregnancy and breastfeeding manuals is that the issue is too overwhelming. Pregnant women are vulnerable; mothers are vulnerable, and so *overrun* with responsibility. In writing about the toxic pollutants they're passing to their unborn children, am I not simply adding to their anxiety?

But I am bothered by this matter of responsibility. Public health campaigns touching on the issue of environmental exposures have focused almost exclusively on smoking, alcohol and drug use – and in some ways this makes sense, since these are substances that, with luck and information and sufficient support, we can at a personal level bring under our control. Yet without an awareness of the other chemicals to which we're being exposed, we remain trapped in the closed room of our own personal culpability. We're prevented from being able to identify who *else* is responsible for keeping us, our offspring and the offspring of others safe – and we are unable to hold them to account.

Shamefully perhaps, it was not until I became a mother that I began knotting the threads together: body, baby, environment. It took that split to recognise: I am porous. I am both myself and not myself, more animal and more contaminated than I knew. I had to step towards that image of the natural, idealised mother for her to pop and leave a gap, a space for something new. Which is why, when midway through our Zoom call this researcher tells me about choosing not to have kids, I am moved but not fazed. I respect him; I admire him. He sees the flood too, and more extensively than me. But I do not dissolve into a pool of guilt and fear and crazed hysteria, which is perhaps what he was worried about. Instead, I listen. I keep talking. It is very necessary now to know.

—

I watched the spiders' nests and I watched them, and it seemed to me that the eggs might at any moment burst with the fullness of the bodies inside. They made me cry. Lots of things made me cry during that time: a parcel in the post, the sight of her foot in the sunlight, stories of beluga whales, a sliver of moon. It was the hormones, people said – as though, hormones being of the body, the tears were somehow inferior; as though they were not quite real.

My favourite nest was the one above the shower. This one I studied each morning in the few minutes during which I would reliably be left alone. Undesiring of company, I found myself in company anyway – a spider mother and her clutch of eggs, hanging but a few centimetres from my head.

When they hatched, it happened all at once. There was no infestation. I looked up one morning to find the web strung with emptied cases, and the spiderlings beside them – poised, as though gathering themselves. The next day, they were gone.

―

There is one more principle left. This one is very short. It is the principle of mixing.

EDCs act at very low concentrations, and can work in combination, meaning that where two or more are mixed, their toxicity may be amplified. Very little is known about these cocktails or their effects, yet the chemicals continue to be licensed and manufactured. Knowing this, I'll leave a gap here – a space for things we know too little about. A decision not to infill; my act of resistance.

MOTHER ANIMAL

STOP. STOP GENERATING. MARK THE PLACE WHERE YOUR KNOWLEDGE ENDS.

———

The spider *Stegodyphus lineatus* lays one small clutch of eggs during her whole lifetime. The young are completely helpless when they emerge, and dependent on the mother, who feeds them by regurgitating a liquid 'milk'. But after two weeks, a surprise change occurs: her young consume her in a process known as 'matriphagy' – an instance of suicidal maternal care, since the mother makes no attempt to escape her fate.

Her dissolution begins soon after egg-laying. Cell walls inside her body become blurred, tissues melt into each other.

The process intensifies as offspring emerge from the egg sac. As she feeds them, she has already begun dying; she stops repairing the web, since she no longer needs to eat – the regurgitated feed is composed of a combination of digested nutrients and her own tissues. Just before matriphagy occurs, her abdomen liquifies completely; her ovaries and heart are the last organs to melt away.

Some days, wanting to escape the house, I'd lift the baby into her sling and head out – to the woods, the meadows, along the hedgerows – until I could feel the pulse of my own blood, the heat of my skin, and so be reassured by the return of some kind of edge; a boundary between she and I, between myself and the rest of the world.

When I met people on the lane that summer they no longer commented on the weather, though it was beautiful that year, the trees bursting with green, the meadows engulfed in wildflowers. Neither, after a time, did they comment on the pandemic. Instead, they seemed to want to hone in on the small bundle pinned to my front, encased in a stretch of fabric. *Isn't she tiny! Is she sleeping? Is she good?*

She *was* tiny, and it made me nimble. Higher and higher up the hill, through undergrowth, around thickets. Unthinkingly, I must have left the trace of myself – footprints, scent, snapped twigs – just as, returning home, I'd find my skin marked with scratches, mud stains, burrs.

Did you know, I whispered to her. *Did you know, did you know.* The spider mother is eaten alive and she doesn't resist it. She offers herself. She is destroyed, and in her destruction she makes something new; she makes life.

—

Today, downstairs, my daughter has just come in from outside. My neighbour is here too, with her daughter – they've all been to the playground together.

I can hear her calling. She wants juice, milk, toys, telly. —Now! she demands – excited, forthright. —Now, now!

I'll go down in a moment. First, I want to share a thought I had this morning as we rolled sausages out of salt dough. I was thinking of the experts I've spoken with – the toxicologists and epidemiologists whose research is woefully underfunded and always on the backfoot, since as one chemical is banned three substitutes appear on the market, sometimes more toxic than before. I was thinking of the questions they asked: *What if bans didn't relate to individual substances? What if they encompassed whole chemical groups? What if chemicals were properly tested before they went on the market? What if regulation was truly protective? What if it prioritised bodies and environments over profit?* They are forceful, these researchers. They apply pressure, they stem flows, and there's a beauty to this, and a power. Yet what struck me today, as we sat at the table, sunshine lighting dust motes in the air, was the enormity of their undertaking. It is like a very extensive form of nest-building, of provision. The toxicologist understands that her home is far greater than the space of her house. She sees that it is everywhere and that it belongs to others too, living now and in the future.

Pressure is building. One campaigner I spoke to talked about a 'chemical awakening' – a shift in public consciousness. When I began reading about forever chemicals, few people I knew had heard of them; now, it seems that everyone has. At the time of writing, water companies in England and Wales are required to test for some PFAS, but are only obliged to act if concentrations exceed 'high risk' levels. It's a bar that falls short of more protective approaches taken by Scotland and the EU, and does not factor in the effects of chemical mixtures, nor the claim by some scientists that when it comes to certain chemicals, no 'safe' level of exposure exists. In the meantime, the EU is working on a group-wide restriction on consumer products, while a number of US states have passed laws banning the use of forever chemicals in items ranging from food packaging and cleaning agents to menstrual products and dental floss – a move that has prompted some corporations to simplify their production chains by going completely PFAS-free.

Added to grassroots pressure on manufacturers are the increasing calls from investors and shareholders, concerned by what one reporter described as a 'tsunami' of recent litigation cases brought against manufacturers as the health effects of forever chemicals become recognised. Yet banning chemicals before the point of manufacture is one thing; the question of what to do about the many persistent chemicals already at large in our environment is another. Clean-up operations are difficult and costly. In the USA, many billions of dollars have been spent on removing PCBs from rivers and estuaries, but they are so pervasive, and have accrued in such large quantities, that little can be done to stop their spread completely. Instead, researchers I spoke to emphasised the need for a broader, more holistic approach: by reducing other stressors in marine ecosystems, such as by-catch, prey depletion and overfishing, we can help build the resilience of wildlife to withstand the effects of toxic contaminants, alongside other human-made pressures.

This morning, as she squished the dough with her fist and as I was thinking these things, I found myself wondering if this moment of chemical awakening (of reckoning with our porosity and our potential for harm) might also herald something new. Perhaps what it means to thrive is changing; perhaps so is how we care. And then of course I was thinking of webs again – of threads strung between multiple points, multiple places.

But I must go. She is waiting at the bottom of the stairs – she has remembered I am home. —Now, now! she shouts, at the top of her lungs. Never has it all felt so urgent.

V
NESTS

Four months passed, then five, and then somehow it was September, and she was six months old. The heat lingered long into autumn that year, the shift in season marked instead by a change in the light; a lengthening of the shadows; late afternoons bathed in an uncannily warm low glow. For days, distant tractor engines rumbled around the valley as farmers cut the meadow grass for silage. Soon after, a fresh troop of spiders arrived through the windows – long-legged, carnivorous and horny. They scuttled out from under the sofa and danced bandily across the red stone tiles as the baby lay and babbled on her playmat. The males of some species make gifts of silk-wrapped food items or small pebbles as a precursor to mating; some immobilise the jaws of females as they inject their sperm. Some females sink their fangs into their partner's abdomen and hold on tight.

There is more than one way to spin a web, I learned that summer, and more than one site to place it. Slipping outside in the early mornings, the baby cocooned in thick cloth across my front, I began to spot the different tricks of construction. Loose spirals on the hawthorn bush; dense sheets like a skin over the grass. Down by the back step, a messy tangle of threads plugged a gap where the plaster was coming away from the wall.

There were nests all over, that year. Under logs, inside cupboards, at the tops of trees. We see everywhere the thing that is constantly on our minds, I suppose, and in that first phase of motherhood, not only was I hypersensitive to threat – I also became peculiarly alert to the means by which other creatures carve out their dwelling places, their nooks and niches.

Yet I was not just an observer, looking on. We overlapped, these nest-makers and I. We were sometimes in conflict. In August the ants had returned, spilling from a loose tile in the front doorway. This time I'd abandoned the hoover, instead laying lemon wedges and cinnamon bark around the entrance, having read that this worked as a deterrent. It didn't. After picking three or four scrambling bodies from the baby's crib, I flicked the kettle switch – shocked, then, at how easily my urge to protect had been transmuted into harm.

I watched the baby, watching me. Steam rose from the tiles, and the sharp scent of citrus filled the room. Some inner part of herself, some just-discernible aspect that was wholly her own, was gaining traction. She was able to fix her gaze; to focus, and follow a moving body. There was an app on my phone that tracked her developmental milestones. Apparently she'd grasped that an object could be partially occluded; that things are not always visible in their entirety. She was also capable of perceiving edges. She understood, now, where one surface ended and another began. Thus her merged and soupy world had begun to splinter into a place of distinct objects – of gaps and breaks and discontinuous forms. She was learning to distinguish between things, to separate them out, as I was in the throes of what connected.

—

I sat on our new stain-resistant sofa to feed her, and the lists on my phone grew, and they grew – fringed and ratcheted along by a breathless kind of horror. Sometimes, almost nostalgically now, I remembered the advertisements I used to flick through during pregnancy; the mothers cradling their smiling infants, a crib in the background, a few toys in the crib, and all of it so pure and transparent and simple. How I longed then to be taken in by that nursery, and by its implicit suggestion that with motherhood comes a path to redemption – a turn towards nurturing and naturalness, and away from the distant violences of the modern world. The women were dressed in plain colours, as I recalled. Neutral tones that blended almost with the natural, neutral walls. As though mothering were synonymous with disappearance; as though disappearance was synonymous with good. I wondered what that nursery smelled like. I wondered if it ever housed spiders, ants. Perhaps in truth it was not a real nursery at all, but a mock-up built in a photography studio. Perhaps the sense of enclosing walls, of a window just out of shot, was artifice – perhaps the room did not exist.

One morning, pulling a bright shirt from the wardrobe, a moth flicked up from between the sleeves, leaving its larvae to gnaw ragged holes through fabric that I could not rid of the stench of my own milk. I shook the shirt. Nothing fell. The moth – the mother, the moth – pressed herself flat against the wall. Had she any intimation of what befell the cloth after her larvae hatched? Did she know the damage caused?

As I pulled on the shirt and as the baby blew raspberries from the bed, I remembered this: the common newt wraps her eggs individually in folded pond leaves; the *Sehirus* burrowing bug deposits hers in a dip, a small depression in the soil. (*Imagine that!* I exclaimed to the baby in a mode of exaggerated speech that had always annoyed me, and that I found myself imitating nonetheless. *A dip, a spot of nothing at all, a nest that leaves no trace!*)

I had looked to other animals for stories to help make sense of a change I did not know how to prepare for, but now the stories themselves were changing, reconfiguring themselves. *Imagine this*, I whispered, when we were back downstairs and lying together on the playmat. *Imagine this, imagine this.* The tailorbird stitches leaves with hair-like plant fibres to make its nest. The weaver bird weaves them. The Eurasian penduline tit makes a house like a softly woven pouch, which hangs suspended from its branch. The Gila woodpecker drills holes in spiky cactus trees, prompting the tree (which notes the wound in its trunk) to secrete a sap that hardens over time, guarding against loss of moisture and supplying the nest with a tough outer layer of protection.

She caught her toes and stuffed them in her mouth, and I tickled the backs of her knees. Of course she could not *imagine*.

Of course she did not *know*. Yet still I told her, still there was something in the telling that felt like sharing, like reaching for a world bigger than these walls.

—

Outside, the leaves were turning red-gold. I sat at the table with my boyfriend, watching walkers pass by the window as we shared the last Magnum from the freezer. Another lockdown was due to be announced, and we were contemplating the prospect of a winter spent more or less confined to the house. The room was tense. It had not been a good morning and it had not been a good night – the baby had woken almost hourly, again, and I felt almost translucent with exhaustion.

At four o'clock I'd knocked on the door to the back bedroom and woken him. —Can you take her? I'd asked. —Just for an hour or so. Needling my voice into a sound I hoped was not too pleading, not too deranged with sleeplessness. I had become so panicky, sitting in there with her – a pillow behind me, another under my elbow, being doggedly sucked. Without sleep, I would surely be the cause of some terrible disaster the next day; I'd drop her, burn the house down, make just the kind of terrible mistake one is warned about on motorways. *Tiredness kills.* I could well believe it.

He'd rolled over and got out of bed, and I had been awash with gratitude; whereupon he'd disappeared downstairs. —Where are you going? —I need the loo. —Take her with you? I'd asked, offering her out. Sleep's oblivion was now within my sights and I was so eager, God help me, to be rid of her.

I'd followed him down to the bathroom. —Jesus! he'd said, putting a hand up to bar the doorway. —Give me a minute, can't you? I can't take her when I'm half asleep!

So back up the stairs we'd tramped, she and I, to wait for half an hour as he'd rounded off his ablutions and made a cup of tea and some toast, calling up cheerily to ask if I wanted some as her eyelids finally began to droop and as her body lulled into sleep – at which point he'd appeared in the doorway, mug in one hand, plate in the other. Ready for action. —She's asleep? Why didn't you tell me?

I'd crawled wretchedly into the back bedroom to lie awake, ringing with rage and adrenaline, while the two of them slept soundly on the other side of the wall. Something had dawned on me in those early, alone hours. A difference between him and me, quietly and almost imperceptibly instated. When it came to responding to her needs, I seemed to lack a gap. She cried and I simply responded – no pause, no lag in which the mind might enter and consider things. He, by contrast, seemed to have retained the gap. He expected it. As such, he was still in possession of his needs. He was aware of them. I remembered that luxury.

At the table, he passed me the ice-cream. It wasn't easy for him either, he said. To be secondary, sidelined, like this. He had to look after himself. A friend of his had told him yesterday to enjoy every minute, and it had made him sad, he said, because I wasn't, I didn't seem to be. What was the problem? Why was I not? —I'm so tired, I said, to which he replied —You're always *so tired*, to which I had no answer.

To an extent, I empathised. Tireder and tireder, narrower and narrower were the scraps of myself that I passed him, while each morning I arose and found the reserves within myself to give her more.

Anyway, as I said, the room was tense, and the tension persisted as the ice-cream was finished, and as the stick was left lying on the table for the other one of us to dispose of, melting indiscreetly into the wooden top, until eventually he went out, saying he needed some air, and I let out an exhale, as I used to when he arrived home.

A friend of mine told me a fissure opens in a relationship after a child is born. —You have to build it again, she said. —You have to make it a triad. She waited until after I'd had the baby to share this nugget. Or I don't know, perhaps she'd said it before and I hadn't been listening – anyway, I was unprepared. Anyway also, it was not a crack so much as a rewiring, as though our whole circuit – he, I, this house and now her – ran with a different charge. How unfamiliar he now seemed, under this new voltage. How foreign the feeling when he touched me, and how different the words when he spoke. *Were* they different? Perhaps they were. My accent was changing, he said. And my real laugh had been replaced by a weird laugh – a fake one. Meanwhile he'd grown a ponytail and taken out a membership with a wine wholesaler; a new crate arrived once a month.

The washing machine pinged – another load was finished. The day continued, as days do, the house at once familiar and changed. I played with the baby. I fed her finger-sized vegetable sticks one by one, and one by one she rubbed them into her chin and dropped their remains merrily on the floor. I wiped her face. I crawled around her little toes and loved them, and I mopped up her vegetable goo – caught as I did by a sudden longing to bring back the ants, who would have set upon such a feast, in all their rampant togetherness.

Underground ant nests possess a hidden subterranean architecture, invisible from the crevice entrances we see. Chambers and passageways, sloping ramps and tunnels through which the colony maintains a steady temperature and humidity. It's a mode of existence that nurtures more than just the ants themselves. Tunnels improve soil drainage and aeration, and as fresh organic matter is pulled underground it benefits the wider ecology too. *I take it back!* I wished, I wanted to tell them. *Give me your underground architecture, your powers of recycling, your clever way with warmth!*

The day was almost over. Small victory – though beyond the past few hours, I remembered little of it. Now I scooped up the baby and carried her to the bathroom, her forehead resting in the crook of my neck. Next I undressed her and I bathed her and I patted her with a towel, then I spread her with cream and put on her nappy and got her dressed again, in a babygrow that had once been soft, many moons ago. I smoothed the worn-out fabric. It had passed through the homes of at least three of my friends and, quietly, it connected me with them.

Upstairs I read her stories and I fed her and I put her to sleep, easing down the blinds he'd put up the weekend I went away, effecting a gold-tinged semi-dark. People were out in their gardens. People were enjoying the last of the summer. I lay down beside her and listened to the sounds of kids playing, sprinklers turning, and to the creaks and clunks of the house turning itself over – gently, gently – the cracks infinitesimally opening, the walls imperceptibly pulling apart. Then as the room sunk into dark, finally I took out my phone and began my work; slotting bookmarks into folders, arranging folders into subfolders. Sorting, shuffling, rearranging again.

—

It might have been a week, two weeks later. I was back on the sofa to feed her, scooping my breast back inside my bra with a neat plastic *clip!*, then scooting her over to the other side, where she set to work again. Strange what the mind does on so little sleep, swinging from exhaustion to peculiar forms of clarity. The animal stories kept coming, as though being mechanically dredged – flickering across my line of vision to shadow a pile of laundry, a plate of half-eaten toast, the baby's plumping limbs.

I didn't mind this. Truth be told, I liked it. The animals were dredged, and I spoke them aloud, and something in the speaking soothed us both.

Some of them go round and round, I told her. Jonah's icefish carve circular nests on the seafloor, with walls of fine sediment and a hard, rocky centre. *Some of them are noisy, some are barely nests at all.* Guillemots lay eggs on stony ledges, in loud and crowded colonies. The eggs are conical in shape, meaning they roll around, not along – an evolutionary quirk that makes them less likely to fall off.

There was a bang on the window. I jumped and looked up to see a delivery guy waving at me cheerily. The baby had fallen asleep. Possibly I had been asleep too. I wiped a trail of saliva from my cheek, rearranged my shirt. Spotting the bundle in my arms, his face had creased into a beaming grin.

I wondered, for a moment, what he saw. Did I look natural, at home here? My navy shirt blended with the sofa, but not the walls behind. And did he see the toys and half-drunk coffee cups scattered about my feet, the bag of unpacked shopping by the door? Did he sense the strange vigilance that followed me now from one end of the day to the next – not static, not calm, but jarred?

—Package for you, love! He heaved it up: the wine delivery. I was going to risk waking the baby for the sake of the fucking wine delivery.

Carefully, I began extricating myself, laying her down centimetre by centimetre on the sofa. She didn't stir. Success! I jumped up, putting on a show for him of domestic capability. Find the key, put the key in the door, turn it, step outside, retrieve the box, wait for him to leave.

He still hadn't, though he did take a step back as I appeared in the doorway. People did that then. But, having stepped back, he promptly stepped forward a bit, since we were in Derbyshire, and people are friendly in Derbyshire, even in a global pandemic. —You were out for the count! That beaming grin again.

I fake-laughed – *silly old me!* – as he peeped over my shoulder at the sofa. —She's waking up, he smiled, and then let out a yelp. I turned, shrieked and arrived just in time to catch her as she launched head first off the endocrine-disrupting sofa and over the brain-injuring tiles.

So, she'd learned to roll. New milestone. —Phew! the guy said. —That was close! Then raised his hand (in – what? A salute? A warning?) and was off.

I watched him go. I was shaking, I noticed. Yes, I was very noticeably shaking, so not *at home* – no, in fact definitely not at home after all. The baby grinned and pumped her legs, still energised and enlivened by the discovery that she could move herself in and of a piece. She wanted to do it again, she wanted to roll and roll – gravity was interesting, as was her body, as was it doing things.

I set her on my hip and went back to the doorway, began shifting the box with my foot. A woman was walking up the lane with a golden retriever. —Hello, she said, coming level with the doorway. She stopped a moment, held my gaze. Wellies, raincoat, grey hair, a kind face. —Are you OK?

Fragments of speech, from late last night, flickered and echoed around my head. Part of me wanted to run, another to make a grab for her. —Yes, I said. There was no space that I could see for being anything else.

Pepper, cucumber, courgette, celery. All these things and more littered the floor, and were cleaned up, and thrown down again. Her range grew. Handfuls of oversteamed vegetables were flung with force into the room.

I picked them up and fed them to the compost heap. I bought reusable baby wipes. Cut them up to make more wipes. I removed the rug, a pot plant, from her growing radius. Beneath the rug was a thick layer of dust. Toxic dust. I pulled out the hoover, disappeared the dust. But the dust was still in the machine. So then I unclipped the canister, opened the back door, put a hand over my nose and mouth as I sent the dust tumbling into the wheelie bin. But where was I sending it – where would it go?

At night, up with her, I searched my memory for animal stories that might help with the questions, the muddled feelings, building up around me. How to safeguard what is porous? How to nest without harm? And what to make of the fierce protectiveness I felt that was neither neutral nor calm?

A lace bug will throw herself in front of predators, sacrificing herself so that her young may live. An earwig will ramp up caregiving behaviours when she senses pathogens are present, cleaning her eggs over and over to protect against infection.

I needed complexities, fine skills – ways of thinking ecologically. Anything but passive. Anything but neutral, still. To be handed this electric alertness while maintaining an image of tranquillity was to be disappeared, suffocated.

Now, remembering that time, I find myself thinking of something different – not lace bugs and earwigs, but a walk I took with a friend in a city park, years ago. Climbing a hill, the two of us had become slightly separated, so that I was a little behind and he a little in front, his anoraked form just nearing the crest when suddenly he stopped and crouched down to better peer at something in the grass. —*Look!* he mouthed, turning to hurry me. —*Look!*

A nest! Minutely woven, intricately crafted thing pulled from a tree by a storm the night before. A place, a real place! We looked and we looked. Twigs, dried grasses, lengths of plastic string, strands of human hair and plastic wrappers, a soft cushion of moss.

We searched the ground for eggs, chicks, and found none. What should we do? Could we take it away with us – might I have it as a souvenir, an ornament reminding me of this creature's capacity to form something, a beauty, from what it found?

We did not. Take it away, that is. Instead we lifted it a few paces and placed it beneath a tree, not knowing quite why we felt so careful about it, so compelled to *do* something; except perhaps that it was so very fragile, and so clearly dislodged.

A nest in a tree in a park in a city, layered up with the layers of the place.

The nest was a wonder – but, I see now, a far from simple one. While their generalist approach to nesting means that many bird species have adapted well to urban environments, the incorporation of human-made materials can change a nest's features. It might drain rainwater less effectively, or more so; it might have a different capacity to absorb or retain heat, or be more conspicuous to predators. Since nests are built with the twin aims of providing incubation and protection, these aspects matter. Too warm or too cold and embryonic development may be disrupted; too visible and the nest may be more vulnerable to predators.

Dissections of urban bird nests have revealed plant seeds and mosses swapped for insulation materials and polyester, dry leaves for plastic and paper bags, sticks for electric cables and plastic straws, thorny branches for anti-bird spikes. These substitutions extend to the means by which some birds are defending themselves against nest-dwelling mites. In the wild, house finches self-medicate against mites and ticks by collecting plants that actively repel them; in cities, they've begun replacing this plant matter with used cigarette butts, since nicotine works as a repellent. The birds are improvising; the birds are clever. But cigarettes are also genotoxic, meaning that while the use of nicotine has been shown to improve chick hatching, fledging success and immune response in the short term, the whole family is contaminated with substances that do long-term damage.

This is why it matters so much that we see mothering animals in relation to their wider contexts. They're amazing, these creatures. They're amazing, they're amazing and they're vulnerable.

They have things to tell us, too. Do an internet search for anthropogenic nesting materials and one quickly comes across images of entanglement. Seabirds caught in fishing lines; plastic string wound around matted necks; crooked feet ensnared by brightly coloured wires. They make me nervous, these bodies, strangled or ensnared by their own shelters – by objects that have flowed, unimpeded, through our home.

Clicking away, I come across a story about Madrid, where refuse dumped in landfill has been spilling from municipal sites into surrounding neighbourhoods, as a growing stork population scavenges the open-air pits for nesting materials. So abundant is the site as a source of food as well as nesting that, prompted also by the effects of climate change, hundreds of birds have ceased making the difficult migration to South Africa – the storks, once a symbol of new birth, have become the risky resurrectors of human waste.

―

According to the app on my phone, the baby had learned to perceive depth and distance. She could distinguish between different intensities of smiling, telling a full from a half one. She could see as far as we could.

I held out my palm in front of her – at an arm's length, a room's width.

Can she see this?

Can she see this?

I held out my face and grinned. I smiled hugely, I smiled with the whole of myself, and still I worried I was not smiling enough.

What was one to do with a small baby? Was I supposed to keep her occupied? I hunted through drawers and cupboards for scarves and fabric sewn with sequins and little mirrors that I hung in swathes above her, and that she plunged into her mouth and sucked.

I constructed ornate hanging devices with rattles and shakers for her to grab and pull. I made silly faces, sang silly songs, and she laughed, and she laughed, and at night in my sleep sometimes I saw things falling on her, things infecting her – little limbs trapped in terrible positions, terrible places. Friends told me this was normal; just another evolutionary protection mechanism. But in the dreams, I was always unable to move. I screamed, I yelled, and yet I could do no thing. Was this normal? Was this evolution?

———

I notice, as I write, that there are two images circling. Sometimes this research that I am doing seems like that of a burying beetle – her movement underground, her necessary proximity to decomposition and death. Other times it seems more like that of a spider, hanging suspended as she spools her threads, back and forth, round and round, joining those points in space.

There were other dreams that followed me through that time. Wild dreams, of windswept hilltops – rock, grass, lichen, moss – things that hung around as I struggled out of sleep, steering my body blindly towards hers, still ringing with altitude and exposure.

She awoke to the sound of the dustbin lorry. Still shaking off my dream world, I held her to the window to see the flashing lights, the great masticating machinery, the men who smiled and waved as she flapped her little hands and bobbed her head, her mouth popping like a fish.

Bags and bags of it. Nappies, broken toys, empty food containers, sanitary pads and plastic packets and God knows what else. Watching it go, I imagined it spreading – into landfill, into nests, into other bodies and their infants. Whittling and spreading, whittling and spreading, just as I felt sometimes that I was drifting outward from my body, my nerves surfacing in other things.

The question of what happens to persistent substances after they're released into the environment is not an easy one to answer. We know that plastics break down into smaller and smaller parts, and that as they do they can leach other contaminants that were added or became by-products during the process of manufacture. We know also that they're lipophilic, meaning they can act as a magnet for other chemicals

– a capacity that makes it almost impossible to predict their toxicity, since the same plastic released into two different environments may accumulate different additional pollutants, with different characteristics, at different concentrations.

If little is known about the precise interactions of these substances, or their effects on living organisms, we do know that as they're subsumed into soil, river, air and ocean systems, they follow fairly predictable patterns of movement – and that these movements are global. Which means that a chemical manufactured and disposed of in one country may turn up years or even decades later in another; that what flows into the back of the dustbin lorry may reappear in places distant. 'We are the parents of other times and places,' writes Rebecca Giggs – 'we indirectly set the conditions by which life elsewhere makes its migrations, its abodes, its nests and niches.'

One of the key transport routes for persistent chemicals is via animal bodies. The seabirds whose nests are littered (or crafted) with plastic debris are also unwitting vectors – along with nesting materials, they can carry hidden pollutants that have escaped into marine ecosystems back onto land via their faeces and dead bodies (on islands off the coast of Nova Scotia where seabirds gather in large colonies, one study found levels of selenium and zinc so high that the land exceeded soil safety guidelines).

This is one of the most disturbing aspects for me about the movements of chemical pollutants. Their routes are natural – they're borne on wildlife and water, wind and air. Yet if their flood is global, it is far from evenly spread. Instead, it forms hotspots; it collects in certain places, certain bodies – a fact revealed in the 1980s when Éric Dewailly, a Canadian epidemiologist and medical researcher, visited the Inuit community of Nunavut in search of a group to use as a 'pristine' baseline comparison in a breast milk contamination study. To Dewailly's surprise, the milk of women in Nunavut contained a level of PCBs several times higher than the urban Quebecois women who were supposed to be the subjects of his study, despite the fact that PCBs had never been used or produced in or anywhere near the region. Their milk also contained a number of other pollutants, including pesticides – another surprise finding in an area with so little agriculture.

Among these pesticides were dieldrin and mirex, which – along with PCBs – would go on to be banned some twenty years later as part of the Stockholm Convention's initial 'dirty dozen'. So high were the levels of contamination among Nunavut infants that they came to be known, chillingly, as a 'living test tube' for those studying the effects of breast milk contamination. Rates of meningitis, bronchitis and pneumonia were high; over 80 per cent of breastfed infants had at least one serious ear infection in their first year, and around a quarter suffered chronic hearing loss. The traditional diets of that region relied on the meat of blubber-rich marine mammals – long-lived predators such as seal and whale that can accumulate highly biomagnified contaminants over decades. Aquatic and marine ecosystems tend to have longer food chains than terrestrial ones, so biomagnification among populations with high fish and seafood diets can be extreme. This remote community turned out to be one of the most polluted on the planet.

As with the insight that pollutants are making their way through the placenta and into foetal tissues, the discovery that toxic pollutants are being carried to remote Arctic regions brings about a strange feeling – a breathless, heinous kind of reach. What is this world if nowhere is separate – nowhere safe?

There is a further reason why these legacy contaminants are showing up in polar regions. Many of them are semi-volatile, meaning they evaporate slowly in warmer conditions and condense quickly when temperatures fall. As a result, they rise in tropical and temperate zones and are carried on air currents to cooler latitudes, where they later fall as rain or snow. By this process, they are drawn towards polar regions by a series of seasonal hops; a 'natural' movement of 'unnatural' chemicals, and one that is being further intensified by climate change, since as the planet warms, evaporation rates are increasing and atmospheric transport is speeding up. But there are other climatic factors, too. Extreme weather events are triggering increased erosion, prompting PCBs held in river sediment and on landfill sites to enter waterways and oceans more quickly; meanwhile, warmer temperatures in polar regions are prompting chemicals that have been locked up in snow and ice to be rereleased into the environment.

These are the chemicals of our past, and ones that will continue to act on bodies and ecosystems long into the future. Their levels remain high in Arctic mammals, and they've been linked with fertility problems in several Arctic species. In polar bears, PCB contamination can result in reduced testosterone levels, smaller testes and a weakened penis bone; in females, they can affect ovulation or reduce the chances of bringing a pregnancy to term. In a creature with one of the slowest reproductive rates of any mammal (a female will mate only five times in her whole lifetime), and one already suffering the effects of habitat loss and climate change, such problems can quickly result in population declines.

And there are new challenges: as more chemicals are being licensed and manufactured, and substitutes found to replace those already banned, novel pollutants are showing up in northern regions, among them further EDCs. Some can now be attributed to local sources, rather than distant ones – a consequence of increasing commercial and industrial development in the region.

Such pollutants are known in the literature as 'chemicals of emerging Arctic concern'. There is a lot of 'emerging concern' when you read around this topic, because many of the substances showing up in our environments today, like the ones in our bloodstreams, are yet to be fully investigated for their biological and ecological effects. Take forever chemicals as an example. Until recently, it was thought that their ultimate resting place was the deep ocean, where they would ultimately be diluted; but recent studies show they're being flushed back into coastal environments via sea spray as waves crash against rocky coastlines – a finding that suggests a more complex relationship at play between marine and terrestrial systems, and that the story of human-made chemicals is not one of gradual dissolution and retreat, but a more disturbing and as yet unfinished tale of reverberations, sudden hammerings and hauntings.

———

One senses these reverberations, these toxic hammerings, in motherhood. In the months after childbirth, I felt more intimate with death – the baby was vulnerable, she existed closer to the edge between surviving or not, and in order to keep her alive I had to be near it too, moulding myself into a being capable of keeping her from it. There *was* something animal about this, and something wild, and it made the notion of chemical persistence all the more alarming. A substance that refuses to change, that resists breaking down; surely this was motherhood's opposite. Matter that is *im*pregnable – a thing that lets nothing in.

I found, then, that I wanted to make a nest for her from tender, friable things. A careful place, that would fall away in time, its parts melting into something else. I wanted also to keep her safe – but I could see no means of instating a boundary between her and the world, and for a while this had made the work of motherhood seem impossible. In time, I saw it differently. How else to begin crafting a nest except through deep and wilful acknowledgement of what is fragile, and what joined?

    I want her to see this as resistance, a sea-change.

We were hanging washing on the line one day, he and I, back when she had learned to roll and roll and was just contemplating crawling. I was working fast, three baby vests and a T-shirt in thirty seconds, while he was standing a little to one side, perhaps matching up the socks. —It needs to change, I was saying, blood beating in my ears, knowing that with this I was marking out a line, instating a boundary at a time when such things often seemed illusory. I had been holding so much in my head, these past months. Thoughts of what escaped and circulated; thoughts of what we passed her.

    —Yes, he said, a rare moment of confluence. —It needs to change.

There were things, I was realising, that we'd never talked about. Aspects of our relationship, and how it held. In the past, it had always seemed that there was so much time – I mean time to work things out, grow, get better, find the right words, the right course of action. Now there was no time. Her arrival had made that work necessary, and made it critical.

The animals flicked through my line of vision, in and out, over and round, fading one moment and sharpening the next. Sometimes I felt they wanted to tell me things. I tried, I tried so hard, to hear them.

Imagine this, they whispered. *Imagine this, imagine this. What if nesting were powerful? What if it were expansive in a good way, a generous one — what if it were radical?*

It is powerful, they went on. *It is.* There are nests that alter the very structure and substrate of a landscape. Through the corrosive action of their own faeces, bats create a complex topography of ripples and bell-holes on the interior surfaces of cave walls; European bee-eater birds drill holes into empty rock faces that later become home to dozens of other species, from rodents and insects to other birds. Termite nests have collapsed whole buildings; boring isopods (small crustaceous creatures possessing four sets of jaws, two sets of antennae and a pair of near-translucent legs for each of their seven body segments) have been found chewing up polystyrene floats, causing wooden sea docks to fall.

Part of me wanted to whoop at hearing that last one – *not so insignificant now!* But here, as elsewhere, I found myself caught out by contradiction. I looked it up: *boring isopods, polystyrene floats.* It was true. Disintegrating, collapsing floats have been recorded in Asia, Australia, Panama and the USA. But it was complicated. Since boring into the floats involves breaking them up, the action is inadvertently creating a new tide of microplastics – the tiny creatures are at once the passive victims and the active generators of this most pervasive form of waste. They're amazing, they're amazing and they're vulnerable.

—

In and out, over and round I go, threading and threading, trying to build a web big enough to capture motherhood, in all its convoluted complexity – spreading, falling, rushing forward, holding back; sifting in and out of bodies, systems, strains of feeling.

In cities, spider webs are offering a new monitoring index for researchers investigating the spread of microplastic pollution. The tiny particles, held suspended in the gossamer threads, provide a glimpse of the plastics drifting through an area. I learned that and felt parts of myself shivering, as though caught momentarily in the light.

A spider web can be a nest. Some species use it to lay their eggs. But it's also a trap – a device for catching things mid-air. A trap can be a nest, then, and a nest can be a trap – a place to enforce stasis, to mummify, to bind a thing to itself.

—Mummy, she says, looking up from the rug. She has been carefully removing every jigsaw piece from every box and collecting them all into a basket she found in the cupboard. —Mummy, Mummy, what are you thinking about? What are you thinking? Will you play with me?

I'm thinking that I don't know if I am the particle, held aloft, or an agent of the glittering web. Perhaps both. Perhaps everything, all at once. Perhaps the web itself.

This morning I read about rats nesting in car engines; wasps nesting in storm window tracks. I cannot tell her such things now without releasing a barrage of questions (—But why? But why, Mummy, why, but what is it for?), so instead we play together. —Shall we put the pieces back into the right boxes? I ask hopefully, sitting down on the rug. —No, she says. And so we don't.

Yesterday, we took a train to the city – there was an exhibition I wanted to take her to by the artist Eva Koťátková, about a giraffe at Prague's Natural History Museum. During the process of preserving its body, toxic gases had been released, causing the temporary closure of the capital's main public square; the centrepiece of the exhibition was a giant sculpture of the giraffe's innards made from quilted cloth that visitors could sit down and roll around on. I knew she'd enjoy that.

We walked from the station – *don't step on the cracks, don't step on the cracks!* – full of elation at having escaped the house, everything about this place a novelty. Soon we reached a bridge overlooking the river; the water ran smooth and straight, below.

—Look! she shouted, stuffing her head between the railings. —Look, a shopping trolley!

Yes, a shopping trolley, looking strangely debased with its wheels in the air, tossed upside down in the water. And around it, a collection of plastic bottles and old drinks cans circled by a strand of river weed.

Urban rivers are among the most anthropogenically modified landscapes on earth; channelisation, dredging, bank stabilisation and pollution have created smooth, fast-flowing currents largely inhospitable to aquatic life. Yet even in these damaged and degraded places opportunities for wildlife recovery have emerged, in the form of human litter.

Bottles, broken plastic and supermarket trolleys all introduce variation in otherwise homogenised environments. While smaller items offer new surfaces and textures, larger items can impede the river's current, altering its flow as a large rock or a fallen tree would, encouraging the deposition of fine sediment and organic matter where plants can find a niche.

Aquatic macroinvertebrates are known to readily colonise litter with empty interiors or rough surfaces, particularly in places with sandy, unstable substrates. This facilitates the creation of distinct communities – more biodiversity has been found in bricks with holes and grooves dumped into rivers than in the rocks that would naturally reside there.

We stood and watched the water flow. Piece by piece, I recalled these things and fitted them to the picture below. In and around the trolley, I imagined sediment deposited, particles of soil and small plant roots accumulating; further species joining that had not been drawn by the litter itself. I wondered what was missing from my picture – what complexities existed beyond my understanding, what dangers passed invisibly beneath the surface. Still, as I looked, I felt a kind of hope – even a damaged place may recover itself.

The world changes, and keeps on changing. Life keeps looking for somewhere to root.

Over the bridge, she flung: gravel, pigeon feathers, a stick, an acorn she'd found in her pocket, more gravel. People passed by. They smiled at her. She represented something to them, I think – naturalness, or innocence, or both. I tried to get her to stop throwing things. —It's probably not a good idea, I suggested. She laughed, and flung some more. —We might hurt the fish, I tried. That did it. And just like that we were off again, the river forgotten, the water behind us, the city ahead.

VI
COMMUNITIES OF CARE

We're in the water. We're in the water, and the water is blue – the kind of blue one spots from plane windows, a vivid aquamarine, neatly boxed.

We're in the water because it's hot. It's midsummer, in the south of France, and records are breaking again. A few hundred miles south of here, wildfires are raking through sun-baked villages. Last night a journalist photographed a man in a scuba mask flushing flames from his house with a garden hose; locals said they knew something was wrong long before they smelled the smoke, because of the animals, who were behaving strangely.

This is privilege, then – this pool, this pleasure, a group of friends not together like this since before lockdown, before babies, before life turned upside down. At three and a half, my daughter is having her first holiday abroad. Hot chocolate and croissants for breakfast. Paddling in the river, a growing collection of water-smoothed pebbles, friends with space to share.

We are submerged, we come up for air. The water (which contains a mixture of sanitisers, water balancers, clarifiers and algaecides) laps at the sides, is trodden into the wooden decking, burned off by the sun. Mating flies buzz heavily across its surface. —Are they caddisflies? someone asks, vaguely. No one knows. The couplings appear driverless, too weighty. They plough blindly into our heads, our bodies, and they nosedive into the pool. Occasionally we reach for a large fishing net with which to scoop them out, and soon the perimeter is scattered with their dripping, drying partnerships.

The children have begun shivering. They love the water, but they no longer belong to it. Still, they refuse to get out. Eventually, someone thinks to offer snacks: peaches, blueberries, slices of melon – fruits I ate in pregnancy to feel connected with her. It works. The children pile out, reeking of chlorine and sun lotion. My daughter grabs a fistful of whatever's on offer, then turns around to show me. —*Look!*

She grins and I cocoon her in a towel, pull her towards me. A friend beside us, sunbathing behind dark glasses, sits up. He nods at the sky, the mountains in the far distance. —The weather'll break tonight.

Earlier, when we arrived here, the rest of them were already in the pool. We changed quickly and I lowered us into the water, her arms clinging tightly to my neck, her legs winding themselves around my waist. A friend of hers had just learned to doggy-paddle. We waded towards her, my vision half-obscured by the inflatable orange armbands. She was turning in circles, figures of eight. My daughter watched, a moment of confusion, then consternation, clouding her face. Suddenly, she turned to me – frustrated, defiant. —Get off, Mummy! Get off, get off!

And just like that, she was swimming. She was swimming, and she was smiling, and she was laughing.

How I worried, in those early months, that this time might never come; that she might never pull away, never feel compelled to strike out alone. She just seemed so *attached*. Now I shadowed her as she doggy-paddled from one end of the pool to the other, at a loss as to what might be needed from me beyond a distant kind of supervision. Tentatively, I swam a few widths. I circled her. —Are you OK? I asked. —Yes, yes! I'm swimming, I'm swimming!

I laughed. I felt toolless, and weirdly light. Around me, other parents were drifting lazily, their bodies glistening like giant water bugs. —Why don't we do this all the time? One of them laughed – meaning, I think, parenting together, parenting as part of a group. I lifted my feet. I floated. What was I to do without her? And yet what joy, in that moment, to be part of this species – this curiously extended period of dependence, in which each infinitesimal movement outward can prove so visible, and so momentous.

—

It was on the tip of my tongue to tell them, my friends afloat on the water's surface, that caddisflies are nocturnal, therefore an unlikely match for whatever species blundered noisily over our heads. Had I shared this, I might have added some detail, large to me, about how the caddisfly provides almost nothing in the way of parental care. The female selects a good site in which to lay her eggs. That's it, that's the entirety of her labour – a likely momentary sensing of air, water, light and waiting hazard. Once hatched, her larvae spin themselves protective casings with silk strung from their salivary glands. To these, they might add gravel, sand, twigs, shreds of bitten-off plants, fragments of rock, seeds, mollusc shells and – sometimes – plastics (in the process of protecting themselves, the grubs have become one of the first known freshwater generators of microplastics). Such additions help strengthen the case, which shields them as they pick their way along slow-moving stream beds, around stones, behind tree roots. When the time comes to pupate, some larvae simply seal the back and front of their decorated case; it becomes their cocoon.

Stories like this fascinated me, at the tail end of my daughter's first year, as she was learning to stand; to hold my hands and toddle forward; to shout for things she wanted. The offspring born independent, unneeding – I thought them extreme, unthinkable almost – but in fact, in the creaturely world, parental care is relatively rare. Caregiving is costly, it demands energy and attention, leaving parents vulnerable to hunger and predation, and delays further reproduction, reducing the chances of genes being passed on. Even among species that do birth dependent young, some circumvent the task of parenting altogether by laying eggs in the nests of other species.

We carry a particular image of animal mothers, and in mammals (for whom lactation usually imposes a necessary proximity) it's true that females are usually the primary caregivers – but in the wider animal kingdom, parenting happens all ways. Take fish as an example. Most fish don't provide care for their offspring at all, but some do, and among these it is usually the male, not the female, that takes on the task. The three-spined stickleback builds his nest and defends it against predators, fanning his clutch with water to circulate oxygen. The specialist role seems to be motivated in part by a desire to mate: the more eggs he has in his nest, the more likely he is to be selected again by a fertile female (essentially, parenting prowess may be functioning as a measure of sexual attractiveness). The male smooth guardian frog, another solo father, is rather different. He shows little interest in mating, eating or even moving as he guards his clutch, leaving the females of the species to take on the more aggressive role of competing to mate with him. (The eggs hatch after around ten days, whereupon the emergent tadpoles climb onto his back to be ferried to water.)

How we look at other animals is shaped in part by what we expect to find; what we see is bounded by what we know. Until a few decades ago, so strong was the image of the mother-caregiver that researchers might have missed the parenting habits of the male smooth guardian frog altogether. Early ornithologists mistook the male jacana bird, whom they had observed building the nest, incubating the eggs and providing the entirety of parental care to chicks, for female. In fact, the female (who weighs around 60 per cent more than the male) was somewhere else entirely, mating with multiple partners and laying successive clutches, which she then left the males to raise. (If it sounds like she has it easy, there is little enviable about the female jacana's role. Laying repeatedly through the mating season is an extreme adaptation, thought to have evolved to offset the high rate of egg loss to predators among this species – it pushes her to the very limits of physical endurance.)

Alongside devoted animal mothers, then, there are devoted fathers – and there are parents that split the caregiving between them. My old friend the burying beetle shares the task with her mating partner, though if the nest is threatened by predators, it is she who is more likely to aggressively defend it. Among birds, around 90 per cent of species parent in couples (jacanas are a notable exception), from nest-building and incubation to fetching food and feeding it to newly hatched chicks. In some species, particularly where the sex ratio of a population is skewed, individuals have been found to form same-sex pairs: male–male black swan pairs will mate with a female, then incubate the egg she leaves behind (one study found that fledging success was higher among these pairs than opposite-sex couples). On the island of Oahu in Hawaii, where the local Laysan albatross population is disproportionately female, a third of breeding pairs are same-sex females (they tend to raise fewer chicks overall than male–female pairs, but it's a better tactic in terms of population survival than not raising offspring at all). Such partnerships can demonstrate a striking flexibility when it comes to parental roles. Captive female flamingo pairs have been observed dividing tasks as an opposite-sex couple would, with one parent spending more time away from the nest, while the other appears to be assigned the task of looking after the eggs.

Biparental care occurs in around 10 per cent of mammal species, usually among those that form pair bonds. The South American marmoset will pull away from her twin infants a few weeks after birth, at which point the father and any older siblings take on caregiving duties. This shared approach is thought to have evolved due to the high energetic costs of birth and gestation in the species; unlike other mammals, where lactation stalls ovulation, the marmoset is capable of conceiving again shortly after birth – so by the time she ceases to care for her infants, she is likely to be pregnant again.

I learned these things in the afternoons on days when I left the house to give the two of them – him, her – some time alone. When I left the house to *get* some time alone. I'd clamber into the car and drive out to my little studio, where for a couple of hours the animals that had been circling my head would be given space to move.

I was still holding that well-worn image of the animal mother up against other species; still hunting down exceptions, still looking to undermine her, as though I couldn't quite let her go. It helped to think about flamingos, about smooth guardian frogs and fish. It helped to have some distance, a slight remove; it made my mind feel clearer.

I was reading about animals, but of course I was thinking about humans too. I was thinking of the dissonance experienced by a new parent as she steps from her fantasy of motherhood into its lived reality. At first, it isn't disillusionment that she feels; it's dysfunction. She isn't stupid, she carries a healthy degree of scepticism about her, she can usually detect when she's being fed a line – but here, it seems, she's missed something. She absorbed certain ideas about how this would go; took them into herself and termed them 'natural'. As a result, when the dissonance kicks in, her first response is to think that she must be at fault. The Mother: steady, instinctive, self-sacrificing, devoted. Her reality, a little under the surface: uncertain, inexpert, falling blindly in love, a thing part-animal and part-machine. Why does she not feel more *natural*?

She is exhausted, and busy (*self-sacrificing*). She is preoccupied with her infant: what they might need, how to stop them choking or suffocating or starving to death or being exposed to invisible, manifold pollutants (*devoted*). She is doing OK. Still, there are moments when what is under the surface breaks through, and then she is left to unpick what it is about all this that feels so difficult – why the task seems so impossible. Is it normal, she wonders, to think this way? The world seems so hostile suddenly – so full of minute, almost indescribable unsafeties. Was it always like this, or is motherhood tampering with something – some basic ability of hers to tell fact from fiction? Her partner wonders if she is going mad. The health visitor (who for a time is the only other adult she sees apart from the neighbours with whom she exchanges pleasantries) calls it 'baby brain'.

Our own species evolved to parent cooperatively, did you know that? A rarity among mammals, we branched sideways from most other primates and began parenting in groups. *Why don't we do this all the time?* my friend said, dangling his legs over the wet sides of the pool. But we did – we do still, in many parts of the world. Some researchers believe our species' extended period of parental care evolved to facilitate all kinds of opportunities for social learning; skills and abilities that can't be hardwired, that only come through the kind of sustained attention and attachment and socialisation that parenting makes possible. Some believe the menopause evolved so that grandmothers would be around to help.

―

—Stop running around doing everything, he said to me one morning, a Saturday. —But I would stop if I knew you were going to do these things. —But I would do them if you let me.

Round and round we went, in eddying little circles like this. Driving at grooves, working up habits. It was petty, I thought, and yet it wasn't. I heaved a heap of dirty laundry from the bedroom floor, bundled it down the stairs. The two of them were sitting up in bed, reading stories. They looked cosy. They looked like how I had imagined motherhood would be.

I hopped down into the kitchen. There were about ten minutes left before she'd want to feed again, when he'd get up and head for the shower, and then I'd be doing all this one-handed. I stirred the pan of porridge cooking on the stove and turned to the sink: began deconstructing the tower of crockery on the draining board.

Lately, it seemed to me that a tip had set in with regards to our parenting. The tip was not how I'd expected this to go. The tip was a slip, as though someone, some utter dick, had lifted a corner of carpet between he and I, tilting everything that had been placed upon it towards me. Since she breastfed, I spent more time with her; and since I spent more time with her, I was learning to better read and predict her needs, to juggle tasks, to take on more things. There was feedback, a loop – and meanwhile, out of familiarity and habit, I suppose, she had singled me out: *milk, primary handler* – a selection she tended to express with gusto if ever I passed her to someone else.

A friend of mine, recognising this tip, had used the word 'organic'. It'll happen *organically*, she'd said, unless you work very consciously against it. Did she mean 'natural'? But I felt caught, conditioned. Either way, I had expected some choice. I'd envisaged this conscious work, this inorganic activity, as bonding: a formula, *our* formula, a new blend. Another half-baked fantasy. Instead, in truth, the set-up was running on ahead of us – opening fault lines, exposing weaknesses, levering the floor.

Dinner plates, saucepans, mugs. Up they clattered onto

shelves, into cupboards, as a wail emanated from upstairs. My system jangled, hotwired. She was a few minutes before time, but there was no helping it. I tripped back up the stairs, and on spotting me in the doorway her wails reached a pitch. Game over. I levered her from his arms and crawled into bed, leaning backwards into the pillows and the coming oxytocin hit as he disappeared in the direction of the shower.

Sometimes, at such times, I felt angry. Sometimes I felt envious. What was it he had that I lacked that enabled him to brush the tide of waiting tasks aside? What was it that so primed me to take on everything? But perhaps this was the trouble. Perhaps herein lay the tricky bind, the difficult paradox of it all. I gave and I gave and in my giving (which I considered benevolent, which I understood was 'good') lay the very elements I wished to keep her from. Gender norms, plasticisers, forever chemicals, the patriarchy – aspects I had imbibed over the years, aspects I'd fought against; matter I'd accrued from my own mother, as she'd accrued from hers – and back, and back, and far back.

From between the folds of duvet a small fist lifted into the air, the arm at a vertical, prompting a break in my line of thought. The fist waved around a bit. I lifted my own hand and held it, her fist, in my own. I thought it wanted holding, containment, but instead it pulled, broke free and began circling again. One eye was open. She watched me – unhurried, calm. The fist thudded against my chest. *Thud, thud* – and then its fingers splayed: *We're here. We're here, we're here, we're here.*

As her young begin weaning, the orangutan initially prepares food by grinding it up with her teeth, then passing it to her offspring to chew. As they mature, she adapts this technique; they learn to eat whole foods. Her labour is precise, fine-tuned – though no less so, I'd wager, than the caddisfly searching out a surface on which to lay her eggs, or the turtle digging a pit on a beach at just the right depth to achieve a safe incubation temperature, before heading back out to sea. Each occupies a niche; each works within its own strain of expertise and specialism, a mesh of parent, offspring and world.

One morning, while the baby napped, I hoovered up the spider webs. Stuck the nose of the machine into every corner in the house and watched as they whipped up, deep into the plastic cylinder. The spiders, by that time, were long gone. They'd disappeared at the start of winter, leaving their threads hanging empty from the walls.

That evening, the three of us sat together at the table; her eating, me eating, him not. —I hoovered up, I said, gesturing to the ceilings. —Yeah? he said, not lifting his head.

A wrench in the stomach: his suffering. I had wanted so much to help and I was not helping – no, I was definitely not helping at all. Now I felt foolish for even bringing it up. What had I wanted, a medal? Or to describe how it had been to see the webs up close, thickened with dust, their numbers multiplying the higher I climbed? They had lost their shape in the months since the spiders' departure. The threads hung loose, they had suffered from lack of care.

The baby kicked her legs, bashed her spoon against her bowl. With the webs gone, the ceilings were louder somehow – symbols of edges, disappearance. The spider, who is eaten alive by her own offspring; the spider, who has already begun dying before they've even been born.

She jabbed a chunk of sweet potato in the direction of her mouth and missed, smeared it messily across her chin. There was something breathless, for me, about these misfires. I wanted her to make them without risk, forever: to protect her from the dawning of her own self-consciousness, the onset of shame.

Next up was a slice of tomato: bull's eye. She gagged. I jerked forward.

—Don't, he said. —You'll make her anxious.

I drew my hands back, shifted my chair closer. Her round eyes followed me, followed us both. I had taken instruction from books, from blogs, on how to do this. Cut the vegetables just longer than the length of her fist, steam until soft but not too soft. Make them harder, smaller, as she grows. I wondered if he knew how calm we were, could be, when it was just she and I. The gagging reflex is strong, the books said. The gagging reflex keeps them safe.

A piece of courgette sailed through the air and hit the wall beside me with a satisfying smack, leaving a thick and greenish streak. I reached for a cloth and began dabbing at it, cheap white paint dissolving under my fingers to reveal the yellowish magnolia underneath. Layer upon layer; nest upon nest upon nest. I put the cloth back on the table, returned to thoughts of niches.

The orangutan provides the entirety of her infant's food and transport for the first years of his life. She might continue nursing until he's six or seven. Raising him is so resource-heavy that she will leave eight to ten years between births, and has only three or four offspring during her whole lifetime. She gives and she gives – the quintessential devoted mother – but to consider her passive is to misunderstand her. This intensive form of parenting grants her unparalleled access to the next generation; she defines many of their experiences, she shapes their skills and habits. Seen from this perspective, she possesses considerable evolutionary influence – visible in the way a female bird selects a mating partner, or a panda opts to abort her pregnancy on the basis of subtle physiological or environmental cues. Her influence is evident also in a recent study of bonobos, where researchers observed high-ranking mothers managing their son's sexual encounters – chaperoning them to promising partners, and shielding the mating couple from competitors (males who lived with their mothers were three times more likely to father offspring than those who didn't). It seems the active, virile male of old may in some cases have been following stage directions from his mother.

Signalling, signalling, even to myself, even in my own head. Why was I so set upon indirection? A mothering creature is active, she's significant; I understood this, I could see it, but I found that I was reluctant to call her powerful. She's pushed to the limits of physical endurance, sucker-punched with responsibility – she has influence, but she is fallible. Far from the traditional image of the animal mother who knows instinctively what to do, study after study shows considerable individual differences in parenting styles among female caregivers of the same species, and even within a single individual across time. Cooke describes how among first-time baboon mothers, the infant mortality rate can be as high as 60 per cent – it's much lower with the secondborn. Baboons have a strong social hierarchy, and low-ranking females, who must be constantly on guard against attacks by other group members, tend to have a more restrictive and energetically demanding parenting style. Meanwhile, higher-ranking females can take a more relaxed approach – they trust that their infants are safe. The mother's social context affects her well-being, and her well-being affects her parenting; this, in turn, affects the well-being of her offspring. Yet there does appear to be a way for low-ranking females to change their fate. Cooke describes research by evolutionary biologist Jeanne Altmann that found that social bonds between mothers functioned as an important protective factor against stress – and they improved the outcomes of infants, too. Through these friendships, mothers gained the protection of other females and knowledge about nearby food sources, as well as physical comfort (mutual grooming releases endorphins that help dial down the stress response, making for calmer, happier parents). Ultimately, motherhood is a learning process, and that learning is dynamic, and it happens always in the context of her wider community and world. She needs an environment that is safe enough, and supportive enough, to raise her young successfully.

A tractor rolled up the lane, and the windows rattled. Across the table, we'd found some common ground – trading lists of jobs, making plans for the weekend. Perhaps we'd have a walk somewhere. Perhaps we'd see some friends. The baby babbled along with us, pointing to the ceilings, laughing at the walls. I had not anticipated what this small being would do to us. Nothing could be perceived in the way that it had been – I, we, my home, this world, was differently exposed. As I'd been thinking of other animals, I was being seen by this wild, knowing pair of eyes.

—

The friend with the dark sunglasses was right; the weather will break tonight. Clouds soak the hills in the far distance. We're here for a wedding, though it is the least weddingy wedding I have ever attended. It's a gathering – an exercise in community. It is eating, talking, cooking, playing together, and it is watching the storm come in.

The air has cooled. The landscape is quieter. Perhaps it's the crickets. Yes, the sound of crickets has receded and the sky seems heavier somehow; the earth waiting, the grass still.

People are fidgety. The weather forecasters are predicting something big. We're feeling ill-equipped, foreigners again in this place. A couple next to me begin googling where's safest in a thunderstorm. Apparently a car is good, a moving vehicle. Apparently the rubber helps.

In London the heatwave is still going strong. Someone reads a story about a woman making ice lollies for her baby using breast milk. People laugh. A fly falls and pirouettes at my feet.

I stand up, leaving the group to go inside and check on my daughter. The house glows – the sky is darkening outside, and someone's put a lamp on. The kids are sitting in a circle. The oldest has brought out a set of face paints and they are each having a go. Through the window I see men running out to the marquee, climbing onto tables, unplugging thick snarls of electric cable. There's a sound of thunder, distant.

Sitting down on the floor, I take my turn as supervising adult. Around the circle there's a tiger face, a unicorn face and a dinosaur face – beasts remote, mythical and extinct. My daughter climbs onto my lap. —What's this? I ask, pointing to a pink smudge of paint on her chin. —A butterfly. —And this? I ask, seeing a light trail of green over her forehead. —The baby. She curls towards me, curls into me. I want it never to end.

Beside us, a group of friends are standing and chatting. Two mums with a baby, a dad with a toddler whose mum is somewhere over by the pool. I realise I've begun identifying all people with kids in this way: a parent, a parent, a parent. I try again: a woman with a spotted headband who laughs often; another in a denim jumpsuit who is quieter; a man in shorts and broken sandals from whose arm a toddler dangles intermittently. They seem to know each other. They're talking about parenthood, the early months, and as they speak it is like listening to people coming up for air. I am not sure I've heard people talk so openly about it before – how intense it was, how total the change. Like flooding, they say. Like deluge. Is this how the unspoken becomes spoken, I wonder – through these rushes, cascades, of words? I don't know, their speech shifts around. It's fluid, uninhibited. They use natural tropes often: *instinctive, wild, animal*. But the animal they describe is not the mute and passive kind. It's strange, contested, tender, connected – a force both nurturing and, yes, powerful; both unpredictable and magnificent.

They laugh. They seem unselfconscious, unafraid. Perhaps they are not. Either way, it is big for me to hear these things – to hear them spoken. Panic and terror as well as joy, lightened and given ground by being shared.

The toddler needs to pee. His father leads him off in the direction of the bathroom as one of the women, the one in the jumpsuit, smiles at me, aware that I've been listening in. We exchange a few words, and then I notice that she is scanning the group of assembled adults, as though wanting to match me to someone. —Are you with . . .?

I hesitate. —No, I say, hoping she won't need an explanation. Instead she blinks, nods, then sits down beside me, joining the face painting.

In the circle we are called upon to paint a zebra face, then a monkey face, then a monster face – and then my daughter has a hold of my hand and is tugging me back outside.

It's getting dark. The swimming pool is empty. Bats flick low over the surface. I wonder if they sense the storm coming, the electricity in the air. I wonder if my daughter does. Behind us, the woman in the jumpsuit is organising a game. People are laughing, calling out to each other. Something happens, I realise, when we come together like this. Something fearsome, if you still hold to the image of the neatly contained nursery; the mute and passive mother. And then, before I've even finished the thought, my mind is caught on another association. *Bats!* I chase after the memory, a disappearing shadow.

—

The bat is the only mammal capable of true flight. Alongside sending her skywards, her wings, which extend from her greatly elongated forelegs, may be used to scoop insects, or form a raincoat or a sleeping bag, or cradle a sleeping infant. Most bat species are highly social; when seeking out roosting sites, their echolocation calls function additionally as a kind of public broadcast to recruit others in the vicinity. Offspring are raised in all-female roosts, which may be mostly unrelated by blood ties. Allonursing is common – and indeed allomaternal behaviours extend far beyond feeding. Infants are supervised in a communal crèche while their mothers go out hunting; should a pup fall to the ground during this time, he'll emit a loud isolation call, and once an allomother has found him she'll stay with him for up to half an hour, warding off attacks from any out-group females until his own mother arrives and transports him back to the roost.

Maternal caregiving is not necessarily limited to biological kin, then – and, as in the context of birth and allonursing, it may be elicited in females who have never themselves given birth. It can also extend beyond the needs of infants. Provided she's well fed, the common vampire bat, who regurgitates food for her own and others' young, will deliver food to adults in need; one study found that while adult-to-adult feeding usually occurs among biological relatives, if an individual is at risk of starvation and a kin donor is unavailable, non-kin feeders will step in. For this species, a wider safety net exists beyond the immediate family group.

If parental care is rare in the animal world, communal breeding is rarer still. Yet it has been observed in fish, insects, arachnids, amphibians, birds and mammals (and in the context of human evolution, it is critical). Lions raise offspring together. So do honeybees, some mice and killer whales. Like bats, emperor penguins group their young in communal crèches while they go out hunting; so do giraffes, who have been observed sending out distress signals following the death of another's calf. Magpie jays raise their young together – a parenting approach that seems to bring benefits in terms of skills acquisition, since the young birds are better at harvesting arthropods than species raised by fewer helpers.

In some species, the sacrifice of alloparents is extreme. In an orb weaver colony, only 40 per cent of females reproduce, but virgin females nevertheless participate in all aspects of brood care, including matriphagy – that is, they are eaten alive by their own (non-biological) young.

At first glance, alloparenting looks like an evolutionary glitch. It may put an individual at greater risk from predators and will expend precious energy; they could be sacrificing their own reproductive potential. Yet it can also offer hidden benefits. Caring for others' offspring might help a young adult acquire parenting skills that will later come in useful when they have offspring of their own; if the group is linked by biological ties, they may be raising their brother or sister, meaning they're still involved in extending the family line.

Communal breeding is more likely to evolve when the costs to helpers are low, or where the overall productivity of breeders is increased, and in species where offspring remain in the family group past sexual maturity. Climatic factors also play a role; regions with uncertain rainfall or high aridity are both associated with the development of cooperative breeding (in birds and mammals respectively). Resource scarcity in these regions is thought to lead to suppressed reproduction, which then prompts the emergence of alloparental care – a finding that suggests the phenomenon may be increasingly common in years to come.

I was isolated that first year of my daughter's life, but I was not *totally* isolated. There were friends who messaged every week, without fail. Our hypnobirthing group set up a chat on WhatsApp, and someone was always awake at two or four or five in the morning, ready to offer humour, or encouragement, or comfort. Conversations could no longer happen in the way that they had (attentions were broken, our time was not our own), but they could take different forms. I learned to use voice notes, finding I could record and listen to them while out pushing the buggy or walking in the woods. The exchanges were unhurried, receptive, and they could span days or weeks.

A community existed, beyond the walls of our house, and it was important – perhaps it was crucial. I liked the thought that we might still contain the blueprint for a more cooperative approach to raising children, but of course it's rarely a good idea to go looking for templates in the natural world. Nature can be brutal. For the meerkat, a creature that in the harsh, dry climate of southern Africa would be unlikely to survive alone, cooperative breeding is elicited by way of bullying and coercion. Meerkats form fierce matriarchal societies, and the dominant female will attack subordinates who become pregnant, expelling them from the group or even killing their offspring as a means of freeing up helpers to raise her own pups. She might have three or four litters over the course of the year, while her subordinates will be lucky to have one. For this species, caring for another's offspring is simply a means of staying alive, affording a subordinate female the protection of the group – and the notion of the purely benevolent animal mother is a misnomer.

There came a point when I stopped holding other species up against a particular image of motherhood, and tried instead to let them follow their own forms. The mother faded, at least in the way I'd pictured her. The stories grew increasingly peculiar.

The *Publilia* treehopper stays with her young until ants discover the group, when her role as caregiver is over; the task is transferred to the capable ants, who feed on a sugary secretion produced by the nymphs, and defend them against predators. The Alcon blue butterfly lays her eggs on the rare marsh gentian plant and leaves them; once hatched, the caterpillar develops an outer coating with a smell like a particular species of ant larvae, which attracts female ants and prompts them to treat him as one of their own. They might go on caring for him for a couple of years, and will feed him even at the expense of their own offspring. That's unless the wasp species *Ichneumon eumerus* comes along. Should this wasp detect the presence of the caterpillar, she'll enter the ant colony and lay her eggs inside his body; when he forms a chrysalis, the eggs will hatch and the larvae consume his pupating body from inside.

The categories blurred, the roles became indistinct. It was difficult to tell who was parent, in this web of relations, and who infant; who protector, and who in need of protection. Clicking around for more examples of interspecies parenting, it wasn't long before I came across accounts of animal adoption. The ducklings raised by a cat; the lamb raised by a sheepdog; the orphaned chimp cared for by an elderly aunt. This was the stuff of gaudy videos on YouTube, their titles gushing and sensationalised (*Adorable cat cuddles kittens! Orphaned chimp just wants to be held!*). I sent a few to my friend Ana, the collector of live-streamed labours. She replied with a link to an article about a Great Dane raising an abandoned fawn and a smiley face.

I wondered what gave these stories their mass appeal. The helpless infant, rescued against the odds – his needs specific, finite and ultimately met by the good (the *super*) mother. But there was something unsettling to me about the way they'd been plated up for viewers. The mother who mothers against the tide, outside the box of her own species; the mother turned spectacle, turned outrageous; the mother who mothers beyond reason. Was there something freakish for us, for our culture, about the thought of a mother whose care extends beyond her own?

I was sitting at my studio desk. Outside, clouds blanketed the sky. The windows were edged with black mould, and the panes were heavy with condensation. I blinked at the screen on my laptop, struggling to focus.

The mother: steady, instinctive, self-sacrificing, devoted.

The mother: evolving, connected, contaminated, vulnerable – an individual in an uncertain time. How do our starting points alter what we look for; what we find?

My breasts were leaking. Two dark blobs shaded the green of my jumper, the smell tart and thick. I needed to leave, to get back and feed her. I picked up my bag, snapped the laptop closed. In the corridor, someone had stacked a pile of boxes and a large mirror against the wall. Squeezing past, I caught a glimpse of myself, wan and weary, in the glass. I looked old, and odd, my face still darted with the melasma I had not noticed until recently, so rarely did I look at my reflection these days. It made me look owl-like, beastly. I thought of Carl Linnaeus and the assumption implicit in his taxonomy that with a womb and lactating breasts a woman is closer to beasts than rational, thinking man. Fuck it, I thought, maybe I *am* more animal – but I haven't lost my mind.

As I hurried down the stairs, there was one animal story that continued to play on my mind, different from the rest for its air of ambiguity. This was a sighting at sea, off the west coast of Iceland – a weakened pilot whale calf swimming in the company of a pod of orcas. What was striking about the group was not just the presence of the calf, but his position in relation to one of the adult females. The two were swimming in echelon formation, a form of aquatic infant-carrying behaviour that allows a calf to be pulled along by the pressure wave created by an adult's larger body. It is the most energy-intensive form of parental care after lactation – the female was effectively swimming *for* the young pilot whale.

Marine ecosystems are cryptic, the veterinary epidemiologist who led the necropsy on Lulu told me. Relationships are non-linear, there are many layers, there is much we don't understand. Perhaps this was why the sight of the orca swimming with the pilot whale had stayed with me – so much about it seemed to defy explanation. Had the orca rescued the pilot calf, or kidnapped him? Had she lost a calf of her own and found a substitute in him, or was this experience of carrying, of holding, entirely new? And what of the calf – did he want to be there? Was he drawn to the female, as she was to him?

That afternoon, sitting under the dripping windows, I'd watched the clip again and again. The whales surfacing and dipping under, the waves choppy, the sea grey. I'd scanned for the calf, squinted hard at that difficult positioning between the two of them, caught on the thought that some hidden formula might exist – some cryptic, under-the-surface spell that could be cast such that one species would make sacrifices so that others, beyond themselves, might live.

The next confirmed sighting of the pod was a year later, and the calf was gone.

We learn about watery worlds through snatched glimpses, the epidemiologist said, and through what they deliver to us – through what is washed up on beaches. As I drove home that afternoon it occurred to me that we learn about life in the womb in much the same way. I remembered asking the epidemiologist if cetacean mothering might be changing in response to an increasingly uncertain and unstable world – if this wider context might be influencing the form and expression of parental care.

He replied that reproduction is as essential as eating and breathing; it, more than any other aspect of animal behaviour, should be robust. I wondered at the time if he was trying to reassure me. *This is special. This is ring-fenced. This might go untouched.* I was dubious. Perhaps he sensed it. —The truth, he added —is that we don't really know.

The truth is that in the rising sea of human-led change nothing is completely ring-fenced, and little remains untouched.

―

Initially I'd been wary of the surfeit of animals that had arrived into our home. The small mountain of soft toys by her cot accrued elephants and rhinos, lions and tigers, and I found it obscene almost – to simulate an abundance here, within the house, while outside, in the wild, these species' numbers thinned. I wondered if as a new mother I'd been handed more than I realised. Not just my own daughter's needs, but the need of a culture to believe that life would keep thriving, keep multiplying, regardless. (And where else to lose our anxiety about a damaged and uncertain world than in the fabled purity of infancy; the familiar image of the good, 'natural' mother?)

On the baby's changing mat, polar bears danced with rabbits; llamas with arctic foxes. The mixtures were false, the animals mostly remote. *What about creatures in our vicinity?* I'd thought. *What about a mat printed with woodlice; a bib with centipedes?* As time passed, my impression changed. Sometimes these animals, their strange combinations, seemed eerily accurate. Weren't warming temperatures forcing creatures into new and unexpected proximities? Weren't we less remote, and more connected, than we knew? I had read by then about polar bears, forced south by warming temperatures, interbreeding with grizzlies; about coyotes mating with foxes and domestic dogs. Their offspring had been given strange names, too nice on the tongue – the pizzly bear, the coywolf.

Animal parenting *is* changing, of course it is. Seagulls carrying high loads of PCBs are spending more time away from their nests; high levels of mercury have been associated with reduced nesting behaviour and lower involvement by fathers in rearing chicks. The repeated exposure of fish to some pesticides has been linked with nest abandonment.

Such findings don't often make the headlines. They're not about births or deaths, not about extinction or recovery, but something slighter – something present in the detail of interactions between parents and offspring; parents and world.

The changes are not always consistent, or predictable. These are not passive victims, but creatures caught in an active struggle to adapt to a pace of change that would usually be spread across millennia. Under increasingly unpredictable and extreme weather conditions, some usually monogamous female birds are mating with multiple partners, and terminating bonds with males that no longer yield sufficient benefits. Others, either through selection or behavioural plasticity, are ramping up parental care as a buffer against the harsher effects of climate change (though it's worth noting that a similar phenomenon has been observed under unusually stable conditions; crises are not the only precursor of exceptional care).

With warmer springs, seasonal cues are changing and some species are breeding earlier, which can increase population numbers overall. The elusive purple emperor butterfly is emerging two or three weeks earlier than a few decades ago, and mating earlier too; early breeding among red deer on the Isle of Rum has led to females having more offspring over the course of their lifetime.

Yet a shift in the timing or location of reproduction can also mean that the delicate link between mating and the availability of food or nesting sites is lost, making specialists such as the Alcon blue butterfly, who must lay her eggs on the leaves of the marsh gentian plant, particularly vulnerable. For some species, more extreme temperatures can affect life even before the point of hatching. Most birds can partly control the rate of embryonic development through incubation temperature – it's one of the means by which communal breeders are able to roughly synchronise hatching times. Extreme heats will likely compromise this ability, with unknown consequences for fine-tuned systems of collective breeding.

Many reptiles and some fish use incubation temperature in a more specific way, to determine the sex of their offspring. Green turtle eggs buried in warmer sand develop into females; in cooler sand, they grow into males. Currently, the population as a whole is tipping towards females – on current trends, it's estimated that by 2030 just 2.4 per cent will be male. Green turtles can live for seventy years or more; they may not be reproducing fast enough to select for adaptations that would protect them from the effects of climate change – but others are luckier. Some reptiles show a startling degree of behavioural plasticity when it comes to breeding, modifying the timing, location and structure of their nests to increase the chance of their offspring surviving. In warmer temperatures, the eastern three-lined skink is laying earlier in the season; the Australian water dragon nests in shadier locations. Some researchers believe the crocodile's ability to survive mass extinctions may be due in part to their particularly hands-on and flexible approach to parenting: in warmer temperatures, they bury *deeper*.

Mothering creatures evolve to fit the needs of their niche, but those niches are changing, and the extent to which the mothering creatures are able to adapt (rather than remain steady and *un*changing) will be a key factor in determining which species survive into the future.

At home, the spiders were edging in again, threading atop the walls. The baby was still not sleeping, our home was still splitting apart, and though I was learning new forms of resilience, what I wanted sometimes was to be vulnerable – not fearful of what the world harboured, but open to what might be possible within it.

Mothering was strange. It was fluxy, it didn't stay still. I didn't want it to. In fact, I needed it not to; she needed it not to too.

She tottered, unsteadily, on her feet. We moved in slow circles around the house, then went out to the garden – sat and played in the soil, chasing woodlice with our fingers.

To be a mother, it seems to me, is to pick a leaf on which to lay one's eggs; to work as a toxicologist in a lab; to live alert to consequences; to be unseparate; to dig deeper, deeper – to nest in, and through, the dark.

———

I had thought, when I saw the men deconstructing wires from the tent, that the party was ending – in fact, it was just rerouting. Music is starting. Someone is singing; her voice lilts and drifts over the wedding guests, gathering now under canvas. Beyond her, it's dark. She has the lyrics on her phone, her face lit by a bluish glow. Elsewhere, people are dropping tea lights into glass jars, arranging them on tables. The night is huge and smells of insects and dry earth.

My daughter yawns and presses herself against my legs. Suddenly, there's a shout from the house – the clouds are breaking, rain falls. Children rush to the edges of the tent. For moments, for minutes, the sound is deafening. The roof barrels with water, the voice of the singer is lost and the grass turns to mud, to streams. Soon the temptation is too much. My daughter makes a run for it, and a moment later she's out there in the puddles, splashing, splashing, caked in mud and rainwater. Other kids run past her, chasing between tent poles, whooping with each roll of thunder.

I'm thinking half-thoughts: *is she warm enough, is she tired, can we drive in this, should I take her to the toilet before we leave?* And I'm thinking: *it is hard, sometimes, to be a grown-up in this world.* Hard to know the responsibilities and costs of parenting. No wonder our culture seems built to keep us distracted, holding us from an encounter with our consequences until the last possible moment. I'm thinking, and I'm thinking, and then without warning a friend grabs me from behind. Picks me up and carries me over to the edge of the tent, dunks my head under the water pouring from its roof, and now I'm soaking too, and this rain will have particles in it from the things we've made, particles like the ones in my bloodstream, and my daughter's bloodstream, and the bloodstreams of every living thing on this planet – and everywhere is flooding, and the world is turning, and I am both parent and child.

The rain seems to ease for a moment as, in the direction of the pool, there is a splash – someone diving, submerged, and then a delicious pause in which, collectively it seems, we experience the sensation of being underwater, mouth, nose and ears filled with liquid, before there is a shout as they breach the surface, and someone whoops, and the thunder rolls.

ACKNOWLEDGEMENTS

I'm grateful to researchers and chemical experts Amalie Timmermann, Hanna Dusza, Andrew Brownlow and Hannah Evans, who gave up their time to talk to me about their work, sharing what was sometimes very personal material. These conversations moved me so much, and took me far deeper into the world of chemical pollution than I might otherwise have had the courage to go.

Thanks also to the animal experts who were generous enough to answer my questions and offer feedback on early drafts: Claire Robertson, Chris Thouless (via Charles, thank you), Alex Hyde, Mike Benton, James Kinross, Jules Howard. Any inaccuracies that remain are, needless to say, my own.

Gratitude to early readers: Chris (for being so articulate in the company of a noisy toddler), Laura, Kieran, Tom, Camilla and the rest of my writing group. Thanks to Sarah Jackson and the Critical Poetics group at Nottingham Trent University, Lila Matsumoto at the University of Nottingham, Five Leaves Bookshop, Paul Carr and the Haarlem shop for their early encouragement and support. Thanks also to my students at the University of Oxford, who teach me so much about writing, as well as grace.

ACKNOWLEDGEMENTS

Thanks to HMRC for the fifteen hours, and to Arts Council England and the Society of Authors for their generous grant funding. Thanks to my agent Jessica Woollard for her fearless belief in the book, and to my editor Sarah Rigby, with whom it found a true home. Thanks to all at David Higham Associates and Elliott & Thompson, and to cover designer Hayley Warnham.

It is a small miracle to me that this book came into being at all, and special gratitude is owed to the friends and family who supported it, and me, through the writing, and who lent more strength than perhaps they know. Mum, Dad, Nick, Ian, Dulcie, Ed, Naomi, Katy, Laurence, Emi, Gemma and all at Wirksworth playgroup, Bradon and Ran (for the pool!), Kat (for the spells!), Jules, Nancy, Lucy, Liz, Maggie, Alan, Lou and – oh – lovely Asha. You are all such wise, kind folk. The very best.

Niamh: you're a powerhouse. You spurred this book. You made me bold, and keep me humble.

BIBLIOGRAPHY

Chapter I: The Enclosure

Bohannon, C., *Eve: How the Female Body Drove 200 Million Years of Human Evolution* (London: Penguin, 2024)

Capodeanu-Nägler A., Prang M.A., Trumbo S.T., et al., 'Offspring dependence on parental care and the role of parental transfer of oral fluids in burying beetles', *Front Zool*, 15, 33 (2018); doi:10.1186/s12983-018-0278-5

Child: Mother Brain, BBC Radio 4 (accessed 25 July 2024); www.bbc.co.uk/programmes/p0h6wb9r

Cooke, L., *Bitch: A Revolutionary Guide to Sex, Evolution and the Female Animal* (London: Transworld, 2022)

Dalton, J., 'Thousands of chickens die in heatwave at farm supplying major UK supermarkets', *The Independent* (1 Aug 2019); www.independent.co.uk/news/uk/home-news/chicken-uk-heatwave-farm-deaths-lincolnshire-tesco-sainsbury-a9025516.html

Giggs, R., *Fathoms: The World in the Whale* (London: Scribe, 2020)

Lewis, S., *Full Surrogacy Now: Feminism Against Family* (London: Verso, 2019)

Narins, E., 'Adele's name for her pregnancy beard will make you love her even more', *Cosmopolitan* (30 Mar 2016); www.cosmopolitan.com/health-fitness/news/a56006/grow-pregnancy-beard

Noren, S.R., Redfern, J.V., Edwards, E.F., 'Pregnancy is a drag: hydrodynamics, kinematics and performance in pre- and post-parturition bottlenose dolphins (*Tursiops truncatus*)', *J Exp Biol*, 214, 24 (2011), 4151–4159; doi:10.1242/jeb.059121

Sathasivam, R., Selliah, P., Sivalingarajah, R., et al., 'Placental weight and its relationship with the birth weight of term infants and body mass index of the mothers', *J Int Med Res*, 51, 5 (2023); doi:10.1177/03000605231172895

@SloggerVlogger, 'Protective bonobo mother takes care of newborn amid scorching heat at UK zoo', *Newsflare* (24 Jul 2019); www.newsflare.com/

video/305626/protective-bonobo-mother-takes-care-of-newborn-amid-scorching-heat-at-uk-zoo

Steingraber, S., *Having Faith: An Ecologist's Journey to Motherhood* (Cambridge: Da Capo Press, 2001)

Taub, M., Mazar, O., Yovel, Y., 'Pregnancy-related sensory deficits might impair foraging in echolocating bats', *BMC Biol*, 21, 60 (2023); doi:10.1186/s12915-023-01557-7

Thurber, C., Dugas, L.R., Ocobock, C., et al., 'Extreme events reveal an alimentary limit on sustained maximal human energy expenditure', *Sci Adv*, 5, 6 (2019); doi:10.1126/sciadv.aaw0341

Tomita, T., Murakumo, K., Ueda, K., et al., 'Locomotion is not a privilege after birth: ultrasound images of viviparous shark embryos swimming from one uterus to the other', *Ethology*, 125 (2019), 122–126: doi:10.1111/eth.12828

Williams, M., 'Europe swelters in the heatwave – in pictures', *The Guardian* (26 Jun 2019); www.theguardian.com/news/gallery/2019/jun/26/europeans-attempt-to-cope-with-record-heatwave-in-pictures

Chapter II: Birth

Abee, C.R., 'The squirrel monkey in biomedical research', *ILAR J*, 31, 1 (1989), 11–20; doi:10.1093/ilar.31.1.11

Aksel, S., Abee, C.R., 'A pelvimetry method for predicting perinatal mortality in pregnant squirrel monkeys (*Saimiri sciuresus*)', *Lab Anim Sci*, 33, 2 (1983), 165–167

Angier, N., 'Why babies are born facing backward, helpless and chubby', *The New York Times* (23 Jul 1996); www.nytimes.com/1996/07/23/science/why-babies-are-born-facing-backward-helpless-and-chubby.html

Blackburn, D.G., 'Viviparity in reptiles and amphibians', *Encyclopedia of Reproduction*, ed. Skinner, M.K., Second Edition, 6, 443–449 (Academic Press: 2018); doi:10.1016/B978-0-12-809633-8.20645-6

Bohannon, C., *Eve: How the Female Body Drove 200 Million Years of Human Evolution* (London: Penguin, 2024)

Carrell, S., 'Edinburgh zoo's giant panda fails to produce cub', *The Guardian* (11 Sep 2017); www.theguardian.com/world/2017/sep/11/our-giant-panda-tian-tian-not-pregnant-says-edinburgh-zoo

Child: Are They What You Eat, BBC Radio 4 (accessed 23 Feb 2024), 14:45; www.bbc.co.uk/programmes/p0h5h43b

Cooke, L., *Bitch: A Revolutionary Guide to Sex, Evolution and the Female Animal* (London: Transworld, 2022)

Dearden, L., 'Giant panda Ying Ying loses cub after "re-absorbing foetus" at Hong Kong zoo', *The Independent* (8 Oct 2015); www.independent.co.uk/climate-change/news/giant-panda-ying-ying-loses-cub-after-reabsorbing-foetus-at-hong-kong-zoo-a6686526.html

Demuru, E., Ferrari, P.F., Palagi, E., 'Is birth attendance a uniquely human feature? New evidence suggests that bonobo females protect and support the parturient', *Evol Hum Behav*, 39, 5 (2018), 502–510; doi:10.1016/j.evolhumbehav.2018.05.003

Demuth, B., 'On mistaking whales', *Granta*, 157 (2021); www.granta.com/on-mistaking-whales-bathsheba-demuth/

Dial, K.P., Jackson, B.E., 'When hatchlings outperform adults: locomotor development in Australian brush turkeys (*Alectura lathami*, Galliformes)', *Proc Biol Sci*, 278, 1712 (2011), 1610–1616; doi:10.1098/rspb.2010.1984

Diamond, A., 'What bird lays the biggest eggs compared to its body size? Where does "lame duck" come from? And more questions from our readers', *Smithsonian* (Jan 2020); www.smithsonianmag.com/smithsonian-institution/bird-lays-biggest-egg-compared-body-size-origin-lame-duck-180973747/

Diogo, R., Siomava, N., Gitton, Y., 'Development of human limb muscles based on whole-mount immunostaining and the links between ontogeny and evolution,' *Hum Dev*, 146, 20 (2019); doi:10.1242/dev.180349

Divers, J., Williams, D., 'Dystocia (egg-binding) in reptiles', *BHS Bull*, 45 (1993); www.thebhs.org/publications/the-herpetological-bulletin/issue-number-45-autumn-1993/2498-hb045-03/file

Dunsworth, H.M., Warrener, A.G., Deacon, T., et al., 'Metabolic hypothesis for human altriciality', *Proc Natl Acad Sci USA*, 109, 38 (2012), 15212–15216; doi:10.1073/pnas.1205282109

Fenelon, J., 'Some animals pause their own pregnancies, but how they do it is still a mystery', *The Conversation* (30 Oct 2019); www.theconversation.com/some-animals-pause-their-own-pregnancies-but-how-they-do-it-is-still-a-mystery-125635

Flanders-Stepans, M.B., 'Alarming racial differences in maternal mortality', *JPE*, 9, 2 (2000), 50–51; doi:10.1624/105812400X87653.

BIBLIOGRAPHY

Ghosh, P., 'Are animals conscious? How new research is changing minds', *BBC News InDepth* (16 Jun 2024); www.bbc.co.uk/news/articles/cv223z15mpmo

Giggs, R., *Fathoms: The World in the Whale* (Australia: Scribe, 2020)

Hallmann, K., Griebeler, E.M., 'Eggshell types and their evolutionary correlation with life-history strategies in squamates', *PLoS One*, 22, 1 (2015); doi:10.1371/journal.pone.0138785

Hassett, B., *Growing Up Human: The Evolution of Childhood* (London: Bloomsbury Publishing, 2022)

Hogenboom, M., *The Motherhood Complex: The Story of Our Changing Selves* (London: Little, Brown, 2021)

Inzani, E., Marshall, H.H., Thompson, F.J., et al., 'Spontaneous abortion as a response to reproductive conflict in the banded mongoose', *Biol Lett*, 15 (2019); doi:10.1098/rsbl.2019.0529

Jiang, B., He, Y., Elsler, A. et al., 'Extended embryo retention and viviparity in the first amniotes', *Nat Ecol Evol*, 7 (2023), 1131–1140; doi:10.1038/s41559-023-02074-0

Kalinka, A.T., 'How did viviparity originate and evolve? Of conflict, co-option, and cryptic choice', *Bio Essays*, 37, 7 (2015), 721–731; doi:10.1002/bies.201400200

Langley, L., 'Go baby! These animal babies grow up without any help from parents', *National Geographic* (9 Sep 2017); www.nationalgeographic.com/animals/article/go--baby--these-animal-babies-grow-up-without-any-help-from-pare

Li, Y-P., Zhong, T., Huang, Z-H., et al., 'Male and female birth attendance and assistance in a species of non-human primate (*Rhinopithecus bieti*)', *Behav Process*, 181 (2020); doi:10.1016/j.beproc.2020.104248

Limb M., 'Disparity in maternal deaths because of ethnicity is "unacceptable"', *BMJ*, 372, 152 (2021); doi:10.1136/bmj.n152

Lubin, Y., 'Arachnid matriphagy', *Entomology Today* (27 Mar 2015); www.entomologytoday.org/2015/03/27/arachnid-matriphagy-these-spider-mothers-literally-die-for-their-young/

Maag, N., Cozzi, G., Seager, D., et al., 'Dispersal-induced social stress prolongs gestation in wild meerkats', *Biol Lett*, 19, 6 (2023); doi:10.1098/rsbl.2023.0183

Martínez-Burnes, J., Muns, R., Barrios-García, H., et al., 'Parturition in mammals: animal models, pain and distress', *Animals (Basel)*, 11, 10 (2021); doi:10.3390/ani11102960

Merchant, C., *The Death of Nature* (New York: HarperCollins, 2019)

Millman, J., 'Status update: "April" the giraffe is doing well, still pregnant', *NBC New York* (24 Feb 2017); www.nbcnewyork.com/news/local/giraffe-livestream-pregnant-birth-animal-adventure-park-new-york-zoo-watch-calf/403324/

Mortola, J.P., Fisher, J.T., Smith, J.B., et al., 'Onset of respiration in infants delivered by cesarean section', *J Appl Physiol Respir Environ Exerc Physiol*, 52, 3 (1982), 716–724; doi:10.1152/jappl.1982.52.3.716

Motani, R., Jiang, D-y., Tintori, A., et al., 'Terrestrial origin of viviparity in Mesozoic marine reptiles indicated by early Triassic embryonic fossils', *PLoS One*, 9, 2 (2014); doi:10.1371/journal.pone.0088640

Muller, Z., Harris, S., 'A review of the social behaviour of the giraffe *Giraffa camelopardalis*: a misunderstood but socially complex species', *Mamm Rev*, 52, 1 (2021), 1–15; doi:10.1111/mam.12268

Pappas, S., 'Female Komodo dragon saved after her eggs burst', *Live Science* (29 Jun 2016); www.livescience.com/55233-female-komodo-dragon-has-surgery.html

Polar Bear Fact Sheet, *PBS* (9 Dec 2020); www.pbs.org/wnet/nature/blog/polar-bear-fact-sheet/

Pyron, R.A., Burbrink, F.T., 'Early origin of viviparity and multiple reversions to oviparity in squamate reptiles', *Ecol Lett*, 17, 1 (2014), 13–21; doi:10.1111/ele.12168

Roberts, A., *The Incredible Unlikeliness of Being: Evolution and the Making of Us* (London: Quercus Publishing, 2014)

Roberts, E.K., Lu, A., Bergman, T.J., et al., 'A Bruce effect in wild geladas', *Science* 335 (2012), 1222–1225

Singh, G., Archana, G., 'Unraveling the mystery of vernix caseosa', *Indian J Dermatol*, 53, 2 (2008), 54–60; doi:10.4103/0019-5154.41645

Taylor, L., 'Movement behaviour after birth demonstrates precocial abilities of African savannah elephant, *Loxodonta africana*, calves', *Anim Behav*, 187 (2022), 331–353; doi:10.1016/j.anbehav.2022.03.002

Trevathan W., 'Primate pelvic anatomy and implications for birth', *Philos Trans R Soc Lond B Biol Sci*, 370 (2015); doi:10.1098/rstb.2014.0065

Wake, M.H., 'Modes of reproduction verts: hermaphroditism, viviparity, oviparity, ovoviviparity: (general definition with examples)', *Encyclopedia of Reproduction*, ed. Skinner, M.K., Second Edition, 18–22 (Academic Press, 2018); doi:10.1016/B978-0-12-809633-8.20531-1

Wang, D.H., Ran-Ressler, R., St Leger, J. et al., 'Sea lions develop human-like vernix caseosa delivering branched fats and squalene to the GI tract', *Sci Rep*, 8, 7478 (2018); doi:10.1038/s41598-018-25871-1

Watts, H.E., Holekamp, K.E., 'Ecological determinants of survival and reproduction in the spotted hyena', *J Mammal*, 90, 2 (2009), 461–471; doi:10.1644/08-MAMM-A-136.1

White, P.A., 'Maternal rank is not correlated with cub survival in the spotted hyena, *Crocuta crocuta*', *Behav Ecol*, 16, 3 (2005), 606–613; doi:10.1093/beheco/ari033

Whittington, C.M., Van Dyke, J.U., Liang, S.Q.T., et al., 'Understanding the evolution of viviparity using intraspecific variation in reproductive mode and transitional forms of pregnancy', *Biol Rev Camb Philos Soc*, 97, 3 (2022), 1179–1192; doi:10.1111/brv.12836

Wickman, F., 'Is giving birth easier for other animals?', *Slate* (27 Sep 2012); www.slate.com/technology/2012/09/animals-giving-birth-dolphins-bear-newborns-easily-but-hyenas-risk-death.html

Yang, B., Zhang, P., Huang, K. et al., 'Daytime birth and post-birth behaviour of wild *Rhinopithecus roxellana* in the Qinling Mountains of China', *Primates*, 57 (2016), 155–160; doi:10.1007/s10329-015-0506-y

Ziegler, T.E., Sosa, M.E., Colman, R.J., 'Fathering style influences health outcome in common marmoset (*Callithrix jacchus*) offspring', *PLoS One*, 12, 9 (2017); doi:10.1371/journal.pone.0185695

Chapter III: Animal Formulas

Altmann, J., *Baboon Mothers and Infants*, Second Edition (Chicago: University of Chicago Press, 2001)

Ballard, O., Morrow, A.L., 'Human milk composition: nutrients and bioactive factors', *Pediatr Clin North Am*, 60, 1 (2013), 49–74; doi:10.1016/j.pcl.2012.10.002

Bohannon, C., *Eve: How the Female Body Drove 200 Million Years of Human Evolution* (London: Penguin, 2024)

Brulliard, K., 'Why goats used to breastfeed human babies', *Washington Post* (25 Feb 2016); www.washingtonpost.com/news/animalia/wp/2016/02/25/why-goats-used-to-breastfeed-human-babies/

Christiansen, F., Dujon, A.M., Sprogis, K.R., et al., 'Noninvasive unmanned aerial vehicle provides estimates of the energetic cost of reproduction in humpback whales', *Ecosphere*, 7, 10 (2016); doi:10.1002/ecs2.1468

Cooke, L., *Bitch: A Revolutionary Guide to Sex, Evolution and the Female Animal* (London: Transworld, 2022)

BIBLIOGRAPHY

Creel, S.R., 'Reproductive suppression and cooperative breeding in dwarf mongooses, *Helogale parvula*', Unpublished Thesis, Purdue University (accessed May 2024); www.docs.lib.purdue.edu/dissertations/AAI9132436/

Creel, S., Monfort, S., Wildt, D., et al., 'Spontaneous lactation is an adaptive result of pseudopregnancy', *Nature*, 351 (1991), 660–662; doi:10.1038/351660a0

Criswell, R., Crawford, K. A., Bucinca, H., et al., 'Endocrine-disrupting chemicals and breastfeeding duration: a review', *Curr Opin Endocrinol Diabetes Obes*, 27, 6 (2020), 388–395; doi:10.1097/MED.0000000000000577

Dasgupta, S., 'Seven of the most extreme milks in the animal kingdom', *Smithsonian* (14 Sep 2015); www.smithsonianmag.com/science-nature/seven-most-extreme-milks-animal-kingdom-180956588/

Day, N., *Baby Meets World: Suck, Smile, Touch, Toddle: A Journey Through Infancy* (London: St Martin's Press, 2013)

The Editors of Encyclopedia, 'mammal', *Encyclopedia Britannica* (accessed 12 Feb 2024); www.britannica.com/animal/mammal

Engelhardt, S.C., Weladji, R.B., Holand, Ø., et al., 'Allonursing in reindeer, *Rangifer tarandus*: a test of the kin-selection hypothesis', *J Mammal*, 97, 3 (2016), 689–700; doi:10.1093/jmammal/gyw027

Gallo-Reynoso, J.P., Ortiz, C., 'Feral cats steal milk from northern elephant seals', *Therya*, 1, 3 (2010), 207–212; doi:10.12933/therya-10-14

Giggs, R., *Fathoms: The World in the Whale* (Australia: Scribe, 2020)

Gloneková, M., Brandlová, K., Pluháček, J., 'Stealing milk by young and reciprocal mothers: high incidence of allonursing in giraffes, *Giraffa camelopardalis*', *Anim Behav*, 113 (2016), 113–123; doi:10.1016/j.anbehav.2015.11.026

Goldman, J.G., 'Not just mammals: Some spiders nurse their young with milk', *National Geographic* (29 November 2018); www.nationalgeographic.com/animals/article/spiders-nurse-young-with-milk-lactation-arachnids

Hinde, K., 'Richer milk for sons but more milk for daughters: sex-biased investment during lactation varies with maternal life history in rhesus macaques', *Am J Hum Biol*, 21, 4 (2009), 512–519; doi:10.1002/ajhb.20917

Hovenden, F., Janes, L., Kirkup, G., Woodward, K. (eds.), *The Gendered Cyborg: A Reader*, First Edition (London: Routledge, 2000)

Kern, C.C., Townsend, S., Salzmann, A., et al., '*C. elegans* feed yolk to their young in a form of primitive lactation', *Nat Comm*, 12 (2021); doi:10.1038/s41467-021-25821-y

Kinross, J., *Dark Matter: The New Science of the Microbiome* (London: Penguin, 2023)

Lefèvre, C.M., Sharp, J.M., Nicholas, K.R., 'Evolution of lactation: ancient origin and extreme adaptations of the lactation system', *Annu Rev Genomics Hum Genet*, 11 (2010), 219–238; doi:10.1146/annurev-genom-082509-141806

Leung, E.S., Vergara, V., Barrett-Lennard, L.G., 'Allonursing in captive belugas (*Delphinapterus leucas*)', *Zoo Biol*, 29, 5 (2010), 633–637; doi:10.1002/zoo.20295

Lydersen, C., Kovacs, K.M., Hammill, M.O., 'Energetics during nursing and early post-weaning fasting in hooded seal (*Cystophora cristata*) pups from the Gulf of St Lawrence, Canada', *J Comp Physiol B*, 167 (1997), 81–88

Ma, W-C., Denlinger, D., 'Secretory discharge and microflora of milk gland in tsetse flies', *Nature*, 247 (1974), 301–303; doi:10.1038/247301a0

MacLeod, K.J., Lukas, D., 'Revisiting non-offspring nursing: allonursing evolves when the costs are low', *Biol Lett*, 10, 6 (2014); doi:10.1098/rsbl.2014.0378

McClellan, H.L., Miller, S.J., Hartmann, P.E., 'Evolution of lactation: nutrition v. protection with special reference to five mammalian species', *Nutr Res Rev*, 21, 2 (2008), 97–116; doi:10.1017/S0954422408100749

Merchant, C., *The Death of Nature* (New York: HarperCollins, 2019)

Mishra, A., Lai, G.C., Yao, L.J., et al., 'Microbial exposure during early human development primes fetal immune cells', *Cell*, 184, 13 (2021), 3394–3409; doi:10.1016/j.cell.2021.04.039

Mortimer, S., 'A qualitative exploration of the media's influence on UK women's views of breastfeeding', *Br J Midwifery*, 30, 1 (2022), 10–18; doi:10.12968/bjom.2022.30.1.10

Mota-Rojas, D., Marcet-Rius, M., Domínguez-Oliva, A., et al., 'Parental behavior and newborn attachment in birds: life history traits and endocrine responses', *Front Psychol*, 14 (2023); doi:10.3389/fpsyg.2023.1183554

Mota-Rojas, D., Marcet-Rius, M., Freitas-de-Melo, A., et al., 'Allonursing in wild and farm animals: biological and physiological foundations and explanatory hypotheses', *Animals (Basel)*, 11, 11 (2021); doi:10.3390/ani11113092

Notarbartolo, V., Giuffrè, M., Montante, C., et al., 'Composition of human breast milk microbiota and its role in children's health', *Pediatr Gastroenterol Hepatol Nutr*, 25, 3 (2022), 194–210; doi:10.5223/pghn.2022.25.3.194

Oftedal, O.T., 'Lactation in whales and dolphins: evidence of divergence between baleen- and toothed-species', *J Mammary Gland Biol Neoplasia*, 2, 3 (1997), 205–230; doi:10.1023/a:1026328203526

Oftedal, O.T., 'The mammary gland and its origin during synapsid evolution', *J Mammary Gland Biol Neoplasia*, 7, 3 (2002); www.edisciplinas.usp.br/pluginfile.php/5663978/mod_resource/content/1/Oftendal.pdf

Poelman, E.H., Dicke, M., 'Offering offspring as food: *Oophaga pumilio*, a dart-poison frog feeding its tadpoles with unfertilized eggs', *Biol Lett*, 3, 5 (2007), 361–364

Power, M.L., Schulkin, J., 'Maternal regulation of offspring development in mammals is an ancient adaptation tied to lactation', *Appl Transl Genomics*, 2 (2013), 55–63; doi:10.1016/j.atg.2013.06.001

Radbill, S.X., 'The role of animals in infant feeding', *American Folk Medicine: A Symposium*, ed. Hand, W.D. (Oakland: University of California Press, 1976)

Ratsimbazafindranahaka, M., Huetz, C., Andrianarimisa, A., et al., 'Characterizing the suckling behavior by video and 3D-accelerometry in humpback whale calves on a breeding ground', *PeerJ*, 10 (2022); doi:10.7717/peerj.12945

Richter, S., 'Wet-nursing, onanism, and the breast in eighteenth-century Germany', *J Hist Sexuality*, 7, 1 (1996), 1–22; www.jstor.org/stable/3840440

Rogers, F.D., Bales, K.L., 'Mothers, fathers, and others: neural substrates of parental care', *Trends Neurosci*, 42, 8 (2019), 552–562; doi:10.1016/j.tins.2019.05.008

Schiebinger, L., 'Why mammals are called mammals: gender politics in eighteenth-century natural history', *Amer Hist Rev*, 98, 2 (1993), 382–411; doi:10.2307/2166840

Spencer, J., '"The link which unites man with brutes": enlightenment feminism, women and animals', *Intellect Hist Rev*, 22, 3 (2012), 427–444; doi:10.1080/17496977.2012.695194

Steingraber, S., *Having Faith: An Ecologist's Journey to Motherhood* (Cambridge: Da Capo Press, 2001)

Stynoski, J.L., Shelton, G., Stynoski, P., 'Maternally derived chemical defences are an effective deterrent against some predators of poison frog tadpoles (*Oophaga pumilio*)', *Biol Lett*, 10, 5 (2014); doi:10.1098/rsbl.2014.0187

Timmermann, C.A.G., Budtz-Jørgensen, E., Petersen, M.S., et al., 'Shorter duration of breastfeeding at elevated exposures to perfluoroalkyl substances', *Reprod Toxicol*, 68 (2017), 164–170; doi:10.1016/j.reprotox.2016.07.010

Tomaszewska, A., Jeleniewska, A., Porębska, K., et al., 'Immunomodulatory effect of infectious disease of a breastfed child on the cellular composition of breast milk', *Nutrients*, 15, 17 (2023), 3844; doi:10.3390/nu15173844

UPI Archives, 'An experiment to teach gorillas at the Columbus Zoo . . .' (10 Oct 1980); https://www.upi.com/Archives/1980/10/10/An-experiment-to-teach-gorillas-at-the-Columbus-Zoo/4510339998400/

Valenze, D., *Milk: A Local and Global History* (New Haven: Yale University Press, 2011)

Videsen, S.K.A., Bejder, L., Johnson, M., et al., 'High suckling rates and acoustic crypsis of humpback whale neonates maximise potential for mother–calf energy transfer', *Funct Ecol*, 31 (2017), 1561–1573; doi:10.1111/1365-2435.12871

Volk, A.A., 'Human breastfeeding is not automatic: why that's so and what it means for human evolution', *J Soc Evol Cult Psychol*, 3, 4 (2009), 305–314; doi:10.1037/h0099314

Watson, C., Khaled, W.T., 'Mammary development in the embryo and adult: a journey of morphogenesis and commitment', *Development*, 135, 6 (2008), 995–1003; doi:10.1242/dev.005439

Wildman, D.E., Uddin, M., Liu, G., et al., 'Implications of natural selection in shaping 99.4% nonsynonymous DNA identity between humans and chimpanzees: enlarging genus Homo', *Proc Natl Acad Sci USA*, 100, 12 (2003), 7181–7188; doi:10.1073/pnas.1232172100

Wolovich, C.K., Evans, S., French, J.A., 'Dads do not pay for sex but do buy the milk: food sharing and reproduction in owl monkeys (*Aotus spp.*)', *Anim Behav*, 75, 3 (2008), 1155–1163; doi:10.1016/j.anbehav.2007.09.023

Xiang, Z., Fan, P., Chen, H., et al., 'Routine allomaternal nursing in a free-ranging Old World monkey', *Sci Adv*, 5, 2 (2019); doi:10.1126/sciadv.aav0499

Chapter IV: Forever Milk

Anca, A., 'Babies exposed to highly toxic nappies face severe disease threat later in life', *The European Environmental Bureau* (21 Jul 2022); www.eeb.org/babies-exposed-to-highly-toxic-nappies-face-severe-disease-threat-later-in-life/

Andrews, D.Q., Stoiber, T., Temkin, A.M., et al., 'Discussion. Has the human population become a sentinel for the adverse effects of PFAS contamination on wildlife health and endangered species?', *Sci Total Environ*, 901 (2023); doi:10.1016/j.scitotenv.2023.165939

Baltic Sea Centre, Stockholm University, 'Fact sheet: Effects shown of endocrine disrupting chemicals in the marine environment' (Mar 2017); www.su.se/polopoly_fs/1.621107.1659706053!/menu/standard/file/edcs-in-the-marine-environment-fact-sheet-webb.pdf

van Beijsterveldt, I.A.L.P., van Zelst, B.D., de Fluiter, K.S., et al., 'Poly- and perfluoroalkyl substances (PFAS) exposure through infant feeding in early life', *Environ Int*, 164 (2022); doi:10.1016/j.envint.2022.107274

van den Berg, M., Kypke, K., Kotz, A. et al., 'WHO/UNEP global surveys of PCDDs, PCDFs, PCBs and DDTs in human milk and benefit–risk evaluation of breastfeeding', *Arch Toxicol*, 91 (2017), 83–96; doi:10.1007/s00204-016-1802-z

Bernasconi, S., Street, M.E., Iughetti, L., et al., 'Chemical contaminants in breast milk: a brief critical overview', *Glob Pediatr*, 2 (2022), 100017; doi:10.1016/j.gpeds.2022.100017

Biss, E., *On Immunity* (London: Fitzcarraldo Editions, 2015)

Bornehag, C-G., Carlstedt, F., Jönsson, B.A.G., et al., 'Prenatal phthalate exposures and anogenital distance in Swedish boys', *Environ Health Perspect*, 123, 1 (2014), 101–107; doi:10.1289/ehp.1408163

Boztas, S., 'The race to destroy the toxic forever chemicals polluting our world', *The Guardian* (4 Jan 2024); www.theguardian.com/environment/2024/jan/04/the-race-to-destroy-the-toxic-forever-chemicals-polluting-our-world

Burton, J.K., *Napoleon and the Woman Question* (Lubbock: Texas Tech University Press, 2007)

de Cock, M., de Boer, M., Govarts, E. et al., 'Thyroid-stimulating hormone levels in newborns and early life exposure to endocrine-disrupting chemicals: analysis of three European mother–child cohorts', *Pediatr Res*, 82, (2017), 429–437; doi:10.1038/pr.2017.50

Cousins, I.T., DeWitt, J.C., Glüge, J., et al., 'The high persistence of PFAS is sufficient for their management as a chemical class', *Environ Sci Process Impacts*, 22, 12 (2020), 2307–2312; doi:10.1039/d0em00355g

Criswell, R., Crawford, K.A., Bucinca, H., et al., 'Endocrine-disrupting chemicals and breastfeeding duration: A review', *Curr Opin Endocrinol Diabetes Obes*, 27, 6 (2020), 388–395; doi:10.1097/MED.0000000000000577

Dalsager, L., Christensen, N., Halekoh, U., et al., 'Exposure to perfluoroalkyl substances during fetal life and hospitalization for infectious disease in childhood: a study among 1,503 children from the Odense Child Cohort', *Environ Int*, 149 (2021); doi:10.1016/j.envint.2021.106395

De Silva, A.O., Armitage, J.M., Bruton, T.A., et al., 'PFAS exposure pathways for humans and wildlife: a synthesis of current knowledge and key gaps in understanding', *Environ Toxicol Chem*, 40, 3 (2021), 631–657; doi:10.1002/etc.4935

Deane, P., 'A life story of large house spiders', *BBC Autumn Watch* (31 Oct 2012); www.bbc.co.uk/blogs/natureuk/entries/6cb2b54e-8ea4-3d33-b898-7de69d943303

Dusza, H.M., 'Contaminants of emerging concern in the fetal environment: unravelling the exposure and effects of endocrine disrupting compounds and micro(nano)plastics in utero', Unpublished Thesis (2022)

Dusza, H.M., Janssen, E., Kanda, R., et al., 'Method development for effect-directed analysis of endocrine disrupting compounds in human amniotic fluid', *Environ Sci Technol*, 53, 24 (2019), 14649–14659; doi:10.1021/acs.est.9b04255

Elmore, B.J., *Seed Money: Monsanto's Past and Our Food Future* (London: W.W. Norton & Co, 2021)

Endocrine Disruptor Lists, 'Substances identified as endocrine disruptors at EU level' (Jun 2024); www.edlists.org/the-ed-lists/list-i-substances-identified-as-endocrine-disruptors-by-the-eu

Endocrine Society, 'Impact of EDCs on reproductive systems' (accessed Mar 2024); www.endocrine.org/topics/edc/what-edcs-are/common-edcs/reproduction

Environmental Working Group, 'Groundbreaking map shows toxic "forever chemicals" in more than 330 wildlife species' (22 Feb 2023); www.ewg.org/news-insights/news-release/2023/02/groundbreaking-map-shows-toxic-forever-chemicals-more-330

European Chemicals Agency Substance infocard, '4,4′-sulphonyldiphenol' (accessed Mar 2024); www.echa.europa.eu/fr/substance-information/-/substanceinfo/100.001.137

European Chemicals Agency Substance infocard, 'bisphenol B' (accessed Mar 2024); www.echa.europa.eu/fr/substance-information/-/substanceinfo/100.000.933

FIDRA, 'PFAS: The Unwelcome Guests in Our Water Sources' (18 Jun 2024); www.pfasfree.org.uk/news/pfas-in-water-sources

FIDRA, 'Twelve key asks for the chemical strategy' (updated Sep 2023); www.fidra.org.uk/download/12-key-asks-for-the-chemical-strategy/

Gabrielsen, G., 'Levels and effects of persistent organic pollutants in arctic animals', in: *Arctic Alpine Ecosystems and People in A Changing Environment* (2007), 390–412; www.researchgate.net/publication/288232894_Levels_and_effects_of_persistent_organic_pollutants_in_arctic_animals

Garcia, M.A., Liu, R., Nihart, A., et al., 'Quantitation and identification of microplastics accumulation in human placental specimens using pyrolysis gas chromatography mass spectrometry', *Toxicol Sci*, 199, 1 (2024), 81–88; doi:10.1093/toxsci/kfae021

Gaur, N., Dutta, D., Jaiswal, A., et al., 'Role and effect of persistent organic pollutants to our environment and wildlife', in: *Persistent Organic Pollutants (POPs) – Monitoring, Impact and Treatment*, ed. Rashed, M.N. (InTechOpen, 2022); doi:10.5772/intechopen.101617

Goldenberg, S., '"Toxic stew" of chemicals causing male fish to carry eggs in testes', *The Guardian* (21 Apr 2010); www.theguardian.com/environment/2010/apr/21/toxic-stew-chemicals-fish-eggs

Hogue, C., 'Toxic PCBs managed poorly decades after production ceased', *C&EN* (2 Jun 2022); www.cen.acs.org/environment/persistent-pollutants/Toxic-PCBs-managed-poorly-decades/100/i20

Hosea, L., Salvidge, R., '"Forever chemicals" found in drinking water sources across England', *The Guardian* (28 Nov 2023); www.theguardian.com/environment/2023/nov/28/forever-chemicals-found-in-drinking-water-sources-across-england

Iribarne-Durán, L.M., Peinado, F.M., Freire, C., et al., 'Concentrations of bisphenols, parabens, and benzophenones in human breast milk: a systematic review and meta-analysis', *Sci Total Environ*, 806 (Pt 1) (2022); doi:10.1016/j.scitotenv.2021.150437

Jensen, S., 'Report of a new chemical hazard', *New Sci*, 32, 612 (1966); www.exhibits.lib.unc.edu/items/show/7436

Johansen, B.E., 'The Inuit's struggle with dioxins and other organic pollutants', *Am Indian Q*, 26, 3 (2002), 479–490; www.jstor.org/stable/4128495

Knapke, E.T., de P. Magalhaes, D., Dalvie, M.A., et al., 'Environmental and occupational pesticide exposure and human sperm parameters: a Navigation Guide review', *Toxicology*, 465 (2022); doi:10.1016/j.tox.2021.153017

Konkel L., 'Mother's milk and the environment: might chemical exposures impair lactation?', *Environ Health Perspect*, 125, 1 (2017); doi:10.1289/ehp.125-A17

Krook, P., 'Investors with $8 trillion call for phase-out of dangerous "forever chemicals"', *ChemSec* (29 Nov 2022); www.chemsec.org/investors-with-8-trillion-call-for-phase-out-of-dangerous-forever-chemicals/

Kuiper, J.R., Liu, S.H., Lanphear, B.P., et al., 'Estimating effects of longitudinal and cumulative exposure to PFAS mixtures on early adolescent body composition', *Am J Epidemiol*, 2024; doi:10.1093/aje/kwae014

Legler, J., 'Amniotic fluid contains numerous unknown chemicals with hormone like activity', *Utrecht University News* (17 Dec 2019); www.uu.nl/en/news/amniotic-fluid-contains-numerous-unknown-chemicals-with-hormone-like-activity

Li, D., Shi, Y., Yang, L. et al., 'Microplastic release from the degradation of polypropylene feeding bottles during infant formula preparation', *Nat Food*, 1 (2020), 746–754; doi:10.1038/s43016-020-00171-y

Little, B., 'How one bad science headline can echo across the internet', *Smithsonian* (31 Jul 2017); www.smithsonianmag.com/science-nature/how-bad-science-headlines-echo-across-internet-180964259/

Liu, S.H., Feuerstahler, L., Chen, Y., et al., 'Toward advancing precision environmental health: developing a customized exposure burden score to PFAS mixtures to enable equitable comparisons across population subgroups, using mixture item response theory', *Environ Sci Technol*, 57, 46 (2023); doi:10.1021/acs.est.3c00343

Liu, Y., Calafat, A.M., Chen, A., et al., 'Associations of prenatal and postnatal exposure to perfluoroalkyl substances with pubertal development and reproductive hormones in females and males: the HOME study', *Sci Total Environ*, 890 (2023); doi:10.1016/j.scitotenv.2023.164353

Ma, D., Lu, Y., Liang, Y., et al., 'A critical review on transplacental transfer of per- and polyfluoroalkyl substances: prenatal exposure levels, characteristics, and mechanisms', *Environ Sci Technol*, 56, 10 (2022), 6014–6026; doi:10.1021/acs.est.1c01057

Mandard, S., 'French lawmakers vote to ban "forever chemicals" except in cooking utensils', *Le Monde* (5 Apr 2024); www.lemonde.fr/en/environment/article/2024/04/05/french-lawmakers-vote-to-ban-forever-chemicals-except-in-cooking-utensils_6667451_114.html

Marlatt, V.L., Bayen, S., Castaneda-Cortès, D., et al., 'Impacts of endocrine disrupting chemicals on reproduction in wildlife and humans', *Environ Res*, 208 (2022); doi:10.1016/j.envres.2021.112584

Martín-Carrasco, I., Carbonero-Aguilar, P., Dahiri, B., et al., 'Comparison between pollutants found in breast milk and infant formula in the last decade: A review', *Sci Total Environ*, 875 (2023); doi:10.1016/j.scitotenv.2023.162461

Martineau, D., De Guise, S., Fournier, M., et al., 'Pathology and toxicology of beluga whales from the St. Lawrence Estuary, Quebec, Canada. Past, present and future', *Sci Total Environ*, 154, 2–3 (1994), 201–215; doi:10.1016/0048-9697(94)90088-4

Mead, M.N., 'Contaminants in human milk: weighing the risks against the benefits of breastfeeding', *Environ Health Perspect*, 116, 10 (2008); www.ncbi.nlm.nih.gov/pmc/articles/PMC2569122/

Megson, D., Idowu, I.G., Sandau, C.D., 'Is current generation of polychlorinated biphenyls exceeding peak production of the 1970s?', *Sci Total Environ*, 924 (2024); doi:10.1016/j.scitotenv.2024.171436

Melymuk, L., Blumenthal, J., Šáňka, O., et al., 'Persistent problem: global challenges to managing PCBs', *Environ Sci Technol*, 56, 12 (2022), 9029–9040; doi:10.1021/acs.est.2c01204

Mínguez-Alarcón, L., Gaskins, A.J., Meeker, J.D., et al., 'Endocrine-disrupting chemicals and male reproductive health', *Fertil Steril*, 120, 6 (2023), 1138–1149; doi:10.1016/j.fertnstert.2023.10.008

Neslen, A., 'EU unveils plan for "largest ever ban" on dangerous chemicals', *The Guardian* (25 Apr 2022); www.theguardian.com/environment/2022/apr/25/eu-unveils-plan-largest-ever-ban-on-dangerous-chemicals

Nielsen, C., Li, Y., Lewandowski, M., et al., 'Breastfeeding initiation and duration after high exposure to perfluoroalkyl substances through contaminated drinking water: a cohort study from Ronneby, Sweden,' *Environ Res*, 207 (2022); www.researchonline.lshtm.ac.uk/id/eprint/4664553/1/1-s2.0-S0013935121015073-main.pdf

Oehlmann, J., Schulte-Oehlmann, U., Kloas, W., et al., 'A critical analysis of the biological impacts of plasticizers on wildlife', *Philos Trans R Soc Lond B Biol Sci*, 364, 1526 (2009), 2047–2062; doi:10.1098/rstb.2008.0242

Ortiz-Zarragoitia, M., Bizarro, C., Rojo-Bartolomé, I., et al., 'Mugilid fish are sentinels of exposure to endocrine disrupting compounds in coastal and estuarine environments', *Mar Drugs*, 12, 9 (2014), 4756–4782; doi:10.3390/md12094756

Panesar, H.K., Wilson, R.J., Lein, P.J., 'Cellular and molecular mechanisms of PCB developmental neurotoxicity', in: Kostrzewa, R.M. (eds.), *Handbook of Neurotoxicity* (New York: Springer, 2022); doi:10.1007/978-3-030-71519-9_204-1

Persson, L., Carney Almroth, B.M., Collins, C.D., et al., 'Outside the safe operating space of the planetary boundary for novel entities', *Environ Sci Technol*, 56, 3 (2022), 1510–1521; doi:10.1021/acs.est.1c04158

Petersen, K.U., Hærvig, K.K., Hougaard, K.S., et al., 'O-140 Exposure to per- and polyfluoroalkyl substances (PFAS) during pregnancy and male reproductive function in young adulthood', *Occup Environ Med*, 80 (2023); doi:10.1136/OEM-2023-EPICOH.24

Preuss, S., 'Denmark wants to ban PFAS in clothing and shoes', *Fashion United* (3 May 2024); www.fashionunited.uk/news/business/denmark-wants-to-ban-pfas-in-clothing-and-shoes/2024050375418

Ragnarsdóttir, O., Abdallah, M.A-E., Harrad, S., 'Dermal bioavailability of perfluoroalkyl substances using in vitro 3D human skin equivalent models', *Environ Int*, 188 (2024); doi:10.1016/j.envint.2024.108772

Ragusa, A., Notarstefano, V., Svelato, A., et al., 'Raman microspectroscopy detection and characterisation of microplastics in human breastmilk', *Polymers*, 14 (2022); doi:10.3390/polym14132700

Reimann, B., Remy, S., Koppen, G., et al., 'Prenatal exposure to mixtures of per- and polyfluoroalkyl substances and organochlorines affects cognition in adolescence independent of postnatal exposure', *Int J Hyg Environ Health*, 257 (2024); doi:10.1016/j.ijheh.2024.114346

Rich, A.L., Phipps, L.M., Tiwari, S., et al., 'The increasing prevalence in intersex cariation from toxicological dysregulation in fetal reproductive tissue differentiation and development by endocrine-disrupting chemicals', *Environ Health Insights*, 10 (2016); doi:10.4137/EHI.S39825

Salto Mamsen, L., Björvang, R.D., Mucs, D., et al., 'Concentrations of perfluoroalkyl substances (PFASs) in human embryonic and fetal

organs from first, second, and third trimester pregnancies', *Environ Int*, 124 (2019), 482–492; doi:10.1016/j.envint.2019.01.010
Salvidge, R., Hosea, L., 'Revealed: scale of "forever chemical" pollution across UK and Europe', *The Guardian* (23 Feb 2023); www.theguardian.com/environment/2023/feb/23/revealed-scale-of-forever-chemical-pollution-across-uk-and-europe
Salvidge, R., Hosea, L., 'Cancer-causing PCB chemicals still being produced despite 40-year-old ban', *The Guardian* (8 Mar 2024); www.theguardian.com/environment/2024/mar/08/cancer-causing-pcb-chemicals-still-produced-despite-40-year-old-ban
Schiavone, A., Corsolini, S., Tao, L., et al., 'Perfluorinated contaminants in fur seal pups and penguin eggs from South Shetland, Antarctica', *Sci Total Environ*, 407 (2009); doi:10.1016/j.scitotenv.2008.12.058
Sen, P., Fan, Y., Schlezinger, J.J., et al., 'Exposure to environmental toxicants is associated with gut microbiome dysbiosis, insulin resistance and obesity', *Environ Int*, 186 (2024); doi:10.1016/j.envint.2024.108569
Siwakoti, R.C., Cathey, A., Ferguson, K.K., et al., 'Prenatal per- and polyfluoroalkyl substances (PFAS) exposure in relation to preterm birth subtypes and size-for-gestational age in the LIFECODES cohort 2006–2008', *Environ Res*, 237, 2 (2023); doi:10.1016/j.envres.2023.116967
Starnes, H.M., Rock, K.D., Jackson, T.W., et al., 'A critical review and meta-analysis of impacts of per- and polyfluorinated substances on the brain and behaviour', *Front Toxicol*, 4 (2022); doi:10.3389/ftox.2022.881584
Steingraber, S., *Having Faith: An Ecologist's Journey to Motherhood* (Cambridge: Da Capo Press, 2001)
Stohler, S., 'First-in-nation ban on PFAS "forever chemicals" in menstrual products, cleaning ingredients, cookware, and dental floss signed by Minnesota Governor', *Safer States* (19 May 2023); www.saferstates.org/press-room/first-in-nation-ban-on-pfas-forever-chemicals-in-menstrual-products-cleaning-ingredients-cookware-and-dental-floss-signed-by-minnesota-governor-today/
Timmermann, C.A.G., Budtz-Jørgensen, E., Skaalum Petersen, M., et al., 'Shorter duration of breastfeeding at elevated exposures to perfluoroalkyl substances', *Reprod Toxicol*, 68 (2017), 164–170; doi:10.1016/j.reprotox.2016.07.010
United Nations Environment Programme and Secretariat of the Basel, Rotterdam and Stockholm Conventions, 'Chemicals in

plastics – a technical report' (2023); www.unep.org/resources/report/chemicals-plastics-technical-report

Wang, Z., Walker, G.W., Muir, D.C.G., et al., 'Toward a global understanding of chemical pollution: a first comprehensive analysis of national and regional chemical inventories', *Environ Sci Technol*, 54, 5 (2020), 2575–2584; doi:10.1021/acs.est.9b06379

Williams, R.S., Curnick, D.J., Brownlow, A., et al., 'Polychlorinated biphenyls are associated with reduced testes weights in harbour porpoises (*Phocoena phocoena*)', *Environ Int*, 150 (2021); doi:10.1016/j.envint.2020.106303

Wu, Y., Simon, K.L., Best, D.A., et al., 'Novel and legacy per- and polyfluoroalkyl substances in bald eagle eggs from the Great Lakes region', *Environ Pollut*, 260 (2020); doi:10.1016/j.envpol.2019.113811

Chapter V: Nests

Adhem, J., Limb, L., 'Spain's rubbish dumps are attracting white storks but danger lurks in the trash', *EuroNews* (10 Feb 2023); www.euronews.com/green/2023/02/10/spains-rubbish-dumps-are-attracting-white-storks-but-danger-lurks-in-the-trash

Alberts, S.C., 'Social influences on survival and reproduction: insights from a long-term study of wild baboons', *J Anim Ecol*, 88 (2019), 47–66; doi:10.1111/1365-2656.12887

AMAP, 2017, 'AMAP Assessment 2016: chemicals of emerging Arctic concern', Oslo, Norway, xvi+353pp

AMAP, 2018; 'AMAP Assessment 2018: biological effects of contaminants on Arctic wildlife and fish', Oslo, Norway, vii+84pp

AMAP, 2021, 'AMAP Assessment 2020: POPs and chemicals of emerging Arctic concern: influence of climate change', Tromsø, Norway, 16pp

Antonsen, A., 'Research confirms: (insect) moms are the best', *Entomology Today* (11 May 2018); www.entomologytoday.org/2018/05/11/research-confirms-insect-moms-are-the-best/

Banfield-Nwachi, M., '"Nothing but fish nests": huge icefish colony found in Antarctic sea', *The Guardian* (13 Jan 2022); www.theguardian.com/world/2022/jan/13/nothing-but-fish-nests-huge-icefish-colony-found-in-antarctic-sea

Barriquand, L., Bigot, J-Y., Audra, P., et al., 'Caves and bats: morphological impacts and archaeological implications. The Azé prehistoric cave

(Saône-et-Loire, France)', *Geomorphology*, 388 (2021); doi:10.1016/j.geomorph.2021.107785

Bergmann, M., Collard, F., Fabres, J. et al., 'Plastic pollution in the Arctic', *Nat Rev Earth Environ*, 3 (2022), 323–337; doi:10.1038/s43017-022-00279-8

Bond, D., 'Lethal "forever chemicals" taint our food, water and even blood. The EPA is stalling', *The Guardian* (24 Oct 2021); www.theguardian.com/commentisfree/2021/oct/24/pfas-forever-chemicals-epa-pollution

Bradley, J., 'Radioactive waste, baby bottles and Spam: the deep ocean has become a dumping ground', *The Guardian* (12 Mar 2024); www.theguardian.com/environment/2024/mar/12/radioactive-waste-baby-bottles-and-spam-the-deep-ocean-has-become-a-dumping-ground

Carter, G.G., Wilkinson, G.S., 'Social benefits of non-kin food sharing by female vampire bats', *Proc R Soc B* (2015); doi:10.1098/rspb.2015.2524

Catarci Carteny, C., Amato, E.D., Pfeiffer, F. et al., 'Accumulation and release of organic pollutants by conventional and biodegradable microplastics in the marine environment', *Environ Sci Pollut Res*, 30 (2023), 77819–77829; doi:10.1007/s11356-023-27887-1

Chakraborty, P., Chandra, S., Dimmen, M.V., et al., 'Interlinkage between persistent organic pollutants and plastic in the waste management system of India: an overview', *Bull Environ Contam Toxicol*, 109, 6 (2022), 927–936; doi:10.1007/s00128-022-03466-x

Chapuisat, M., Oppliger, A., Magliano, P., Christe, P., 'Wood ants use resin to protect themselves against pathogens', *Proc Biol Sci*, 274 (2007), 2013–2017; doi:10.1098/rspb.2007.0531

Cirino, E., 'More plastic in the world means more plastic in osprey nests', *Audubon* (6 Sep 2017); www.audubon.org/news/more-plastic-world-means-more-plastic-osprey-nests

Commoner, B., Bartlett, P.W., Eisl, H., et al., 'Long-range air transport of dioxin from North American sources to ecologically vulnerable receptors in Nunavut, Arctic Canada: final report to the North American Commission for Environmental Cooperation', *Center for the Biology of Natural Systems*, Queens College, CUNY (2000); www.cec.org/files/documents/publications/2196-long-range-air-transport-dioxin-from-north-american-sources-ecologically-vulnerable-en.pdf

Corrales, J., 'Our garbage, their homes: artificial material as nesting material', *The Nature of Cities* (11 Dec 2016); www.thenatureof

cities.com/2016/12/11/garbage-homes-use-artificial-material-nesting-material-birds/

Cousins, I.T., Johansson, J.H., Salter, M.E., et al., 'Outside the safe operating space of a new planetary boundary for per- and polyfluoroalkyl substances (PFAS)', *Environ Sci Technol*, 56, 16 (2022), 11172–11179; doi:10.1021/acs.est.2c02765

Davidson, T., 'Boring crustaceans damage polystyrene floats under docks polluting marine waters with microplastic', *Mar Pollut Bull*, 64 (2012), 1821–1828; doi:10.1016/j.marpolbul.2012.06.005

Davidson, T., Altieri, A.H., Ruiz, G.M., et al., 'Bioerosion in a changing world: a conceptual framework', *Ecol Lett*, 21 (2018), 422–438; doi:10.1111/ele.12899

Dopico, M., Gómez, A., 'Review of the current state and main sources of dioxins around the world', *J Air Waste Manag Assoc*, 65, 9 (2015), 1033–1049; doi:10.1080/10962247.2015.1058869

Duda, M., Glew, J., Michelutti, N., et al., 'Long-term changes in terrestrial vegetation linked to shifts in a colonial seabird population', *Ecosystems*, 23 (2020); doi:10.1007/s10021-020-00494-8

The Editors of Encyclopedia, 'guillemot', *Encyclopedia Britannica* (accessed 20 Jun 2024); www.britannica.com/animal/guillemot

The Editors of Encyclopedia, 'killer whale', *Encyclopedia Britannica* (accessed 9 May 2024); www.britannica.com/animal/killer-whale

The Editors of Encyclopedia, 'spider', *Encyclopedia Britannica* (accessed 30 Apr 2024); www.britannica.com/animal/spider-arachnid

Elhacham, E., Ben-Uri, L., Grozovski, J., et al., 'Global human-made mass exceeds all living biomass', *Nature*, 588 (2020), 442–444; doi:10.1038/s41586-020-3010-5

Fängström, B., Strid, A., Grandjean, P., et al., 'A retrospective study of PBDEs and PCBs in human milk from the Faroe Islands', *Environ Health*, 4, 12 (2005); doi:10.1186/1476-069X-4-12

Fort, J., Robertson, G.J., Grémillet, D., et al., 'Spatial ecotoxicology: migratory Arctic seabirds are exposed to mercury contamination while overwintering in the Northwest Atlantic', *Environ Sci Technol*, 48, 19 (2014), 11560–11567; doi:10.1021/es504045g

Gabrielsen, G., 'Levels and effects of persistent organic pollutants in arctic animals', in: *Arctic Alpine Ecosystems and People in A Changing Environment*, eds. Ørbaek, J.B., Kallenborn, R., Tombre, I., et al. (Springer, 2007), 377–412; doi:10.1007/978-3-540-48514-8_20

BIBLIOGRAPHY

Gaur, N., Dutta, D., Jaiswal, A., et al., 'Role and effect of persistent organic pollutants to our environment and wildlife', *IntechOpen* (2022); doi:10.5772/intechopen.101617

Giggs, R., *Fathoms: The World in the Whale* (London: Scribe, 2020)

Goßmann, I., Süßmuth, R., Scholz-Böttcher, B.M., 'Plastic in the air?! – spider webs as spatial and temporal mirror for microplastics including tire wear particles in urban air', *Sci Total Environ*, 832 (2022); doi:10.1016/j.scitotenv.2022.155008

Gururaja, K.V., Dinesh, K.P., Priti, H., et al., 'Mud-packing frog: a novel breeding behaviour and parental care in a stream dwelling new species of *Nyctibatrachus* (Amphibia, Anura, Nyctibatrachidae)', *Zootaxa*, 3796, 1 (2014); doi:10.11646/zootaxa.3796.1.2

Hallworth, M.T., Bayne, E., McKinnon, E., et al., 'Habitat loss on the breeding grounds is a major contributor to population declines in a long-distance migratory songbird', *Proc R Soc* (28 Apr 2021); doi:10.1098/rspb.2020.3164

Hendry, L., 'Spider webs: not just for Halloween', *NHM Discover: British Wildlife* (accessed 13 May 2024); www.nhm.ac.uk/discover/spider-webs.html

Hiemstra, A-F., Moeliker, C.W., Gravendeel, B., et al., 'Bird nests made from anti-bird spikes', *Deinsea* (11 Jul 2023); www.hetnatuurhistorisch.nl/fileadmin/user_upload/documents-nmr/Publicaties/Deinsea/Deinsea_21/Deinsea_21_17_25_2023_Hiemstra_et_al.pdf

Higginson, A.D., 'Conflict over non-partitioned resources may explain between-species differences in declines: the anthropogenic competition hypothesis', *Behav Ecol Sociobiol*, 71, 99 (2017); doi:10.1007/s00265-017-2327-z

Hoondert, R.P.J., Ragas, A.M.J., Hendriks, A.J., 'Simulating changes in polar bear subpopulation growth rate due to legacy persistent organic pollutants – temporal and spatial trends', *Sci Total Environ*, 754 (2021); doi:10.1016/j.scitotenv.2020.142380

Jamieson, A.J., Brooks, L.S.R., Reid, W.D.K., et al., 'Microplastics and synthetic particles ingested by deep-sea amphipods in six of the deepest marine ecosystems on Earth', *R Soc Open Sci*, 6 (2019); doi:10.1098/rsos.180667

Jones, L., *Matrescence: On the Metamorphosis of Pregnancy, Childbirth and Motherhood* (London: Allen Lane, 2023)

BIBLIOGRAPHY

Krosofsky, A., 'Plastic toys have a greater impact on the environment and human health than we thought', *Green Matters* (11 Aug 2021); www.greenmatters.com/p/environmental-impact-plastic-toys

Learn, J.R., 'Seabirds bring contaminants to land ecosystems: study', *The Wildlife Society* (29 Jun 2015); www.wildlife.org/seabirds-bring-sea-contaminants-on-to-land-ecosystems-study/

Li, Y.F., Kallenborn, R., Zhang, Z., 'Persistent organic pollutants and chemicals of emerging Arctic concern in the Arctic environment', *Environ Sci Ecotechnol*, 18 (2023); doi:10.1016/j.ese.2023.100332

López-García, A., Aguirre, J.I., 'White storks nest at high densities near landfills changing stork nesting distributions in the last four decades in Central Spain', *Ornithol Appl*, 125, 2 (2023); doi:10.1093/ornithapp/duad009

Mallory, M.L., Mahon, L., Tomlik, M.D., et al., 'Colonial marine birds influence island soil chemistry through biotransport of trace elements', *Water Air Soil Pollut*, 226, 31 (2015); doi:10.1007/s11270-015-2314-9

Marcol, A., Lizana, M., Alvarez, A., et al., 'Egg-wrapping behaviour protects newt embryos from UV radiation', *Anim Behav*, 61 (2001), 639–644; doi:10.1006/anbe.2000.1632

Massart, F., Gherarducci, G., Marchi, B., et al., 'Chemical biomarkers of human breast milk pollution', *Biomark Insights*, 3 (2008), 159–169; doi:10.4137/bmi.s564

McFarland, C., 'Bird nests tell extraordinary stories, if you learn how to read them', *Audubon* (Spring 2023); https://www.audubon.org/magazine/spring-2023/bird-nests-tell-extraordinary-stories-if-you

Montory, M., Habit, E., Fernandez, P., et al., 'Biotransport of persistent organic pollutants in the southern hemisphere by invasive Chinook salmon (*Oncorhynchus tshawytscha*) in the rivers of northern Chilean Patagonia, a UNESCO biosphere reserve', *Environ Int*, 142 (2020); doi:10.1016/j.envint.2020.105803

Morales-McDevitt, M.E., Becanova, J., Blum, A., et al., 'The air that we breathe: neutral and volatile PFAS in indoor air', *Environ Sci Technol Lett*, 8, 10 (2021), 897–902; doi:10.1021/acs.estlett.1c00481

Neslen, A., 'EU unveils plan for "largest ever ban" on dangerous chemicals', *The Guardian* (25 Apr 2022); www.theguardian.com/environment/2022/apr/25/eu-unveils-plan-largest-ever-ban-on-dangerous-chemicals

Novita, N., Amiruddin, H., Ibrahim, H., et al., 'Investigation of termite attack on cultural heritage buildings: a case study in Aceh Province, Indonesia', *Insects*, 11, 6 (2020); doi:10.3390/insects11060385

BIBLIOGRAPHY

Ovenden, T., 'Plastic found lining UK seabird nests on a worrying scale', *The Conversation* (28 Jul 2020); www.theconversation.com/plastic-found-lining-uk-seabird-nests-on-a-worrying-scale-142118

Pavid, K., 'Body snatchers: eaten alive by parasitic wasps', *NHM Discover* (accessed 28 May 2024); www.nhm.ac.uk/discover/body-snatchers-eaten-alive.html

Pelc, C., 'PFAS in rainwater: what it means for health', *Medical News Today* (3 Aug 2022); www.medicalnewstoday.com/articles/pfas-in-rainwater-what-it-means-for-health#How-PFAS-end-up-in-drinking-water

Perkins, T., 'PFAS "forever chemicals" constantly cycle through ground, air and water, study finds', *The Guardian* (18 Dec 2021); www.theguardian.com/environment/2021/dec/17/pfas-forever-chemicals-constantly-cycle-through-ground-air-and-water-study-finds

Perkins, T., 'Alarming levels of PFAS in Norwegian Arctic ice pose new risk to wildlife', *The Guardian* (11 Feb 2023); www.theguardian.com/environment/2023/feb/11/pfas-norwegian-arctic-ice-wildlife-risk-stressor

Poore, G., Bruce, N., 'Global diversity of marine isopods (except asellota and crustacean symbionts)', *PLoS One*, 7 (2012); doi:10.1371/journal.pone.0043529

Renske, P.J., Hoondert, A.M.J., Ragas, A., et al., 'Simulating changes in polar bear subpopulation growth rate due to legacy persistent organic pollutants – temporal and spatial trends', *Sci Total Environ*, 754 (2021); doi:10.1016/j.scitotenv.2020.142380

Routti, H., Atwood, T.C., Bechshoft, T. et al., 'State of knowledge on current exposure, fate and potential health effects of contaminants in polar bears from the circumpolar Arctic', *Sci Total Environ*, 664 (2019), 1063–1083; doi:10.1016/j.scitotenv.2019.02.030

Salvidge, R., 'High levels of toxic chemicals found in Cambridgeshire water supply', *The Guardian* (8 Feb 2022); www.theguardian.com/environment/2022/feb/08/high-levels-of-toxic-chemicals-found-in-cambridgeshire-drinking-water

Sankaran, V., 'Study finds toxic "forever chemicals" may be "intentionally added" to some period products', *The Independent* (11 Aug 2023); www.independent.co.uk/news/science/forever-chemicals-period-products-pfas-b2391356.html

Santamans, A.C., Boluda, R., Picazo, A., et al., 'Soil features in rookeries of Antarctic penguins reveal sea to land biotransport of chemical pollutants', *PLoS One*, 12, 8 (2017); doi:10.1371/journal.pone.0181901

Schæbel, L.K., Bonefeld-Jørgensen, E.C., Vestergaard, H., et al., 'The influence of persistent organic pollutants in the traditional Inuit diet on markers of inflammation', *PLoS One*, 12, 5 (2017); doi:10.1371/journal.pone.0177781

Schmidt, C., 'How PCBs are like grasshoppers', *Environ Sci Technol*, 44, 8 (2010), 2752; doi:10.1021/es100696y

Sha, B., Johansson, J.H., Tunved, P., et al., 'Sea spray aerosol (SSA) as a source of perfluoroalkyl acids (PFAAs) to the atmosphere: field evidence from long-term air monitoring', *Environ Sci Technol*, 56, 1 (2022), 228–238; doi:10.1021/acs.est.1c04277

Shupova, T.V., Koniakin, S.M., Grabovska, T.O., 'Multi-species settlement by secondary hollow-nesting passerine birds in a European bee-eater (*Merops apiaster*) colony', *Ornis Hung*, 30, 1 (2022); 179–188; doi:10.2478/orhu-2022-0014

Soler J.J., Martín-Vivaldi M., Nuhlíčková S., et al., 'Avian sibling cannibalism: hoopoe mothers regularly use their last hatched nestlings to feed older siblings', *Zool Res*, 43, 2 (2022), 265–274; doi:10.24272/j.issn.2095-8137.2021.434

Sonne, C., Dietz, R., Jenssen, B.M., et al., 'Emerging contaminants and biological effects in Arctic wildlife', *Trends Ecol Evol*, 36, 5 (2021), 421–429; doi:10.1016/j.tree.2021.01.007

Steingraber, S., *Having Faith: An Ecologist's Journey to Motherhood* (Cambridge: Da Capo Press, 2001)

Street, S., Jaques, R., De Silva, T.N., 'Convergent evolution of elaborate nests as structural defences in birds', *Proc R Soc B* (2022); doi:10.1098/rspb.2022.1734

Suárez-Rodríguez, M., López-Rull, I., Macías Garcia, C., 'Incorporation of cigarette butts into nests reduces nest ectoparasite load in urban birds: new ingredients for an old recipe', *Biol Lett* (23 Feb 2013); doi:10.1098/rsbl.2012.0931

Suárez-Rodríguez, M., Macías Garcia, C., 'There is no such a thing as a free cigarette; lining nests with discarded butts brings short-term benefits, but causes toxic damage', *J Evol Biol*, 27, 12 (2014), 2719–2726; doi:10.1111/jeb.12531

Suárez-Rodríguez, M., Montero-Montoya, R.D., Macías Garcia, C., 'Anthropogenic nest materials may increase breeding costs for urban birds', *Front Ecol Evol*, 5 (2017); doi:10.3389/fevo.2017.00004

Tallamy, D.W., 'Insect parental care', *BioScience*, 34, 1 (1984) 20–24; doi:10.2307/1309421

Thompson, R.C., Moore, C.J., vom Saal, F.S., Swan, S.H., 'Plastics, the environment and human health: current consensus and future trends', *Philos Trans R Soc Lond B Biol Sci*, 364 (2009), 2153–2166; doi:10.1098/rstb.2009.0053

Turns, A., *Go Toxic Free: Easy and Sustainable Ways to Reduce Chemical Pollution* (London: Michael O'Mara, 2022)

UNEP News and Stories, 'Grasshopper effect serves pollutants onto plates of Arctic peoples' (16 Nov 2018); www.unep.org/news-and-stories/story/grasshopper-effect-serves-pollutants-plates-arctic-peoples

de Valpine, P., Eadie, J.M., 'Conspecific brood parasitism and population dynamics', *Am Nat*, 172, 4 (2008), 547–562; doi:10.1086/590956

Walkinshaw, C., Tolhurst, T.J., Lindeque, P.K. et al., 'Impact of polyester and cotton microfibers on growth and sublethal biomarkers in juvenile mussels', *Microplast Nanoplast*, 3, 5 (2023); doi:10.1186/s43591-023-00052-8

Walton, N., 'The grass-carrying wasp: a solitary wasp that builds an unusual nest', *MSU News* (30 Oct 2023); www.canr.msu.edu/news/the_grass_carrying_wasp_a_solitary_wasp_that_builds_nests_in_unusual_places

Wiesinger, H., Wang, Z., Hellweg, S., 'Deep dive into plastic monomers, additives, and processing aids', *Environ Sci Technol*, 55, 13 (2021), 9339–9351; doi:10.1021/acs.est.1c00976

The Wildlife Trusts, 'Common sexton beetle' (accessed 12 May 2024); www.wildlifetrusts.org/wildlife-explorer/invertebrates/beetles/common-sexton-beetle

Wilson, H.L., Johnson, M.F., Wood, P.J., et al., 'Anthropogenic litter is a novel habitat for aquatic macroinvertebrates in urban rivers', *Freshw Biol*, 66 (2021), 524–534; doi:10.1111/fwb.13657

Wong, F., Hung, H., Dryfhout-Clark, H., et al., 'Time trends of persistent organic pollutants (POPs) and Chemicals of Emerging Arctic Concern (CEAC) in Arctic air from 25 years of monitoring', *Sci Total Environ*, 775 (2021); doi:10.1016/j.scitotenv.2021.145109

Xie, Z., Mi, L., Gandraß, J., et al., 'Final Report: emerging and legacy organic contaminants in the polar regions', *German Environment Agency* (April 2022); www.umweltbundesamt.de/sites/default/files/medien/479/publikationen/texte_46-2022_emerging_and_legacy_organic_contaminants_in_the_polar_regions.pdf

Yang, G., Zhou, W., Qu, W., et al., 'A review of ant nests and their implications for architecture', *Buildings*, 12 (2022); doi:10.3390/buildings 12122225

Zhang, X., Lohmann, R., Sunderland, E. M., 'Poly- and perfluoroalkyl substances in seawater and plankton from the northwestern Atlantic margin', *Environ Sci Technol*, 53, 21 (2019), 12348–12356; doi:10.1021/acs.est.9b03230

Zhong, H., Wu, M., Sonne, C., et al., 'The hidden risk of microplastic-associated pathogens in aquatic environments,' *Eco-Environ Health*, 2, 3 (2023), 142–151; doi:10.1016/j.eehl.2023.07.004

Chapter VI: Communities of Care

Asher, C., 'Climate change is disrupting the birds and the bees', *BBC Future* (8 Aug 2017); www.bbc.com/future/article/20170808-climate-change-is-disrupting-the-birds-and-the-bees

Bales, K.L., 'Parenting in animals', *Curr Opin Psychol*, 15 (2017), 93–98; doi:10.1016/j.copsyc.2017.02.026

Barreca, A., Deschenes, O., Guldi, M., 'Maybe next month? Temperature shocks and dynamic adjustments in birth rates', *Demography*, 55, 4 (2018), 1269–1293; doi:10.1007/s13524-018-0690-7

Bichet, C., Allainé, D., Sauzet, S., Cohas, A., 'Faithful or not: direct and indirect effects of climate on extra-pair paternities in a population of Alpine marmots', *Proc R Soc B*, 283, 1845 (2016); doi:10.1098/rspb.2016.2240

Blackwood, K., 'Prolonged immaturity an evolutionary plus for human babies', *Cornell Chronicle* (8 Feb 2021); www.news.cornell.edu/stories/2021/02/prolonged-immaturity-evolutionary-plus-human-babies

Boisacq, P., De Keuster, M., Prinsen, E., et al., 'Assessment of poly- and perfluoroalkyl substances (PFAS) in commercially available drinking straws using targeted and suspect screening approaches', *Food Addit Contam: Part A*, 40, 9 (2023), 1230–1241; doi:10.1080/19440049.2023.2240908

Boose, K., White, F., Brand, C., et al., 'Infant handling in bonobos (*Pan paniscus*): exploring functional hypotheses and the relationship to oxytocin', *Physiol Behav*, 193, A (2018), 154–166; doi:10.1016/j.physbeh.2018.04.012

BIBLIOGRAPHY

Botero, C.A., Rubenstein, D.R., 'Fluctuating environments, sexual selection and the evolution of flexible mate choice in birds', *PLoS One*, 7, 2 (2012); doi:10.1371/journal.pone.0032311

Brahic, C., 'Parasitic butterflies fool ants with smell', *New Scientist* (3 Jan 2008); www.newscientist.com/article/dn13139-parasitic-butterflies-fool-ants-with-smell/

Breining, G., 'With climate change, species are increasingly interbreeding to survive', *The World* (1 Jun 2015); www.theworld.org/stories/2015-06-02/climate-change-species-are-increasingly-hybridizing-survive

Cant, M.A., 'Cooperative breeding systems', in: Royle, N.J., Smiseth, P.T. (eds.), *The Evolution of Parental Care* (Oxford, 2012; Online Edition, Oxford Academic, 17 Dec 2013); doi:10.1093/acprof:oso/9780199692576.003.0012

Carstens, E.A., 'What's with all the gay penguins?', *Tufts: Museum Studies* (22 Feb 2021); www.sites.tufts.edu/museumstudents/2021/02/22/whats-with-all-the-gay-penguins/

Cecco, L., 'Toronto's mystery predator really is a coy-wolf – but not as we know it', *The Guardian* (8 Apr 2022); www.theguardian.com/environment/2022/apr/08/coy-wolf-hybrid-toronto-canada-aoe

Cooke, L., *Bitch: A Revolutionary Guide to Sex, Evolution and the Female Animal* (London: Transworld, 2022)

Cornwallis, C.K., 'Cooperative breeding and the evolutionary coexistence of helper and nonhelper strategies', *Proc Natl Acad Sci USA*, 115, 8 (2018), 1684–1686; doi:10.1073/pnas.1722395115

Doody, J.S., Guarino, E., Georges, A. et al., 'Nest site choice compensates for climate effects on sex ratios in a lizard with environmental sex determination', *Evol Ecol*, 20 (2006), 307–330; doi:10.1007/s10682-006-0003-2

Du, W.G., Li, S.R., Sun, B-J., et al., 'Can nesting behaviour allow reptiles to adapt to climate change?', *Philos Trans R Soc Lond B Biol Sci*, 378 (2023), 1–14; doi:10.1098/rstb.2022.0153

The Editors of Encyclopedia, 'animal social behaviour', *Encyclopedia Britannica* (accessed 23 Aug 2024); www.britannica.com/topic/animal-social-behaviour

The Editors of Encyclopedia, 'caddisfly', *Encyclopedia Britannica* (accessed 1 February 2024); www.britannica.com/animal/caddisfly

Emlen, S., 'Jacanas and polyandry', *PBS* (2001); www.pbs.org/wgbh/evolution/library/01/6/l_016_04.html

Evans, S., Gustafsson, L., 'Climate change upends selection on ornamentation in a wild bird', *Nat Ecol Evol*, 1, 0039 (2017); doi:10.1038/s41559-016-0039

Fanis, E.D., Jones, G., 'Allomaternal care and recognition between mothers and young in pipistrelle bats (*Pipistrellus pipistrellus*),' *J Zool*, 240(4) (1996), 781–787; doi:10.1111/j.1469-7998.1996.tb05324.x

Faust, K.M., Carouso-Peck, S., Elson, M.R., et al., 'The origins of social knowledge in altricial species', *Annu Rev Psychol*, 2 (2020), 225–246; doi:10.1146/annurev-devpsych-051820-121446

Gaston, A.J., Gilchrist, H.G., Hipfner, J.M., 'Climate change, ice conditions and reproduction in an Arctic nesting marine bird: Brunnich's guillemot (*Uria lomvia* L.)', *J Anim Ecol*, 74 (2005), 832–841; doi:10.1111/j.1365-2656.2005.00982.x

Giggs, R., 'Noiseless messengers', *Emergence Magazine* (27 Jun 2022); www.emergencemagazine.org/essay/noiseless-messengers/

Gloneková, M., 'Maternal behaviour in giraffes (*Giraffa camelopardalis*)', Unpublished PhD Thesis, Faculty of Tropical AgriSciences, Czech University of Life Sciences Prague (2016)

Goldshtein, A., Harten, L., Yovel, Y., 'Mother bats facilitate pup navigation learning', *Curr Biol*, 32, 2 (2022); 350–360; doi:10.1016/j.cub.2021.11.010

Goyes Vallejos, J., Ulmar Grafe, T., Ahmad Sah, H.H. et al., 'Calling behaviour of males and females of a Bornean frog with male parental care and possible sex-role reversal', *Behav Ecol Sociobiol*, 71, 95 (2017); doi:10.1007/s00265-017-2323-3

Groenewoud, F., Clutton-Brock, T., 'Meerkat helpers buffer the detrimental effects of adverse environmental conditions on fecundity, growth and survival', *J Anim Ecol*, 90 (2021), 641–652; doi:10.1111/1365-2656.13396

Hallworth, M.T., Bayne, E., McKinnon, E. et al., 'Habitat loss on the breeding grounds is a major contributor to population declines in a long-distance migratory songbird', *Proc R Soc*, 288, 1949 (2021); doi:10.1098/rspb.2020.3164

Hannam, P., 'Climate change will remove birds' control over hatching eggs: study', *The Sydney Morning Herald* (3 Feb 2016); www.smh.com.au/environment/climate-change/the-early-bird-catches-the-worm-gets-a-new-twist-with-climate-change-study-20160202-gmjjiy.html

Hassett, B., *Growing Up Human: The Evolution of Childhood* (London: Bloomsbury, 2022)

BIBLIOGRAPHY

Horton, H., 'Puffin nesting sites in western Europe could be lost by end of century', *The Guardian* (8 Dec 2022); www.theguardian.com/environment/2022/dec/08/puffin-nesting-sites-western-europe-seabirds-global-heating

Hrdy, S.B., Burkart, J.M., 'The emergence of emotionally modern humans: implications for language and learning', *Phil Trans R Soc B* (2020); doi:10.1098/rstb.2019.0499

Jones, L., *Matrescence: On the Metamorphosis of Pregnancy, Childbirth and Motherhood* (London: Allen Lane, 2023)

Junghanns, A., Holm, C., Fristrup Schou, M., et al., 'Extreme allomaternal care and unequal task participation by unmated females in a cooperatively breeding spider', *Anim Behav*, 132 (2017), 101–107; doi:10.1016/j.anbehav.2017.08.006

Kerth, G., 'Causes and consequences of sociality in bats', *BioScience*, 58, 8 (2008), 737–746; doi:10.1641/B580810

Kluger, J., 'Scientists rush to understand the murderous mamas of the monkey world', *Time* (15 Jun 2011); https://time.com/archive/6934920/scientists-rush-to-understand-the-murderous-mamas-of-the-monkey-world/

Laloë, J.-O., Esteban, N., Berkel, J., et al., 'Sand temperatures for nesting sea turtles in the Caribbean: implications for hatchling sex ratios in the face of climate change', *J Exp Mar Bio Ecol*, 474 (2016), 92–99; doi:10.1016/j.jembe.2015.09.015

Langen, T.A., Vehrencamp, S.L., 'How white-throated magpie-jay helpers contribute during breeding', *Auk*, 116, 1 (1999), 131–140; doi:10.2307/4089460

Leung, E., Vergara, V., Barrett-Lennard, L., 'Allonursing in captive belugas (*Delphinapterus leucas*)', *Zoo Biol*, 29 (2010), 633–637; doi:10.1002/zoo.20295

Lukas, D., Clutton-Brock, T., 'Climate and the distribution of cooperative breeding in mammals', *R Soc Open Sci*, 4 (2017); doi:10.1098/rsos.160897

Mallory, M.L., Mahon, L., Tomlik, M.D. et al., 'Colonial marine birds influence island soil chemistry through biotransport of trace elements', *Water Air Soil Pollut*, 226, 31 (2015); doi:10.1007/s11270-015-2314-9

Martinho-Truswell, A., 'How like the kiwi we are', *Aeon* (11 Apr 2023); www.aeon.co/essays/how-evolution-made-humans-more-like-birds-than-other-mammals

Mateo, J., Cuadrado, M., 'Communal nesting and parental care in Oudri's fan-footed gecko (*Ptyodactylus oudrii*): field and experimental evidence of an adaptive behavior', *J Herpetol*, 46 (2012), 209–212; www.jstor.org/stable/41515039

Møller, A.P., 'Protandry, sexual selection and climate change', *Glob Chang Biol*, 10 (2004), 2028–2035; doi:10.1111/j.1365-2486.2004.00874.x

Mota-Rojas, D., Marcet-Rius, M., Freitas-de-Melo, A., et al., 'Allonursing in wild and farm animals: biological and physiological foundations and explanatory hypotheses', *Animals*, 11, 11 (2021); doi:10.3390/ani11113092

Moyes, K., Nussey D.H., Clements, M.N., et al., 'Advancing breeding phenology in response to environmental change in a wild red deer population', *Glob Chang Biol*, 17 (2011), 2455–2469; doi:10.1111/j.1365-2486.2010.02382.x

Mrusczok, M., Zwamborn, E., von Schmalensee, M., et al., 'First account of apparent alloparental care of a long-finned pilot whale calf (*Globicephala melas*) by a female killer whale (*Orcinus orca*)', *Can J Zool*, 101, 4 (2023), 288–293; doi:10.1139/cjz-2022-0161

Nielsen, J.T., Møller, A.P., 'Effects of food abundance, density and climate change on reproduction in the sparrowhawk *Accipiter nisus*', *Oecologia*, 149 (2006), 505–518; doi:10.1007/s00442-006-0451-y

Nuwer, R., 'Same-sex parenting can be an adaptive advantage', *Smithsonian* (27 Nov 2013); www.smithsonianmag.com/smart-news/same-sex-parenting-can-be-an-adaptive-advantage-180947865/

O'Brien, S.J., Johnson, W.E., Driscoll, C.A., et al., 'Conservation genetics of the cheetah: lessons learned and new opportunities', *J Hered*, 108, 6 (2017), 671–677; doi:10.1093/jhered/esx047

de Oliveira Terceiro, F.E., Burkart, J.M., 'Cooperative breeding', in: Vonk, J., Shackelford, T. (eds.), *Encyclopedia of Animal Cognition and Behavior* (Cham: Springer, 2019); doi:10.1007/978-3-319-47829-6_1351-1

O'Neill, K., 'Bridging binaries in the animal kingdom', *University of Cambridge Museums and Botanic Garden* (11 Jun 2020); www.museums.cam.ac.uk/blog/2020/06/11/bridging-binaries-in-the-animal-kingdom/

Readfearn, G., 'Emperor penguins: thousands of chicks in Antarctica die due to record-low sea ice levels', *The Guardian* (24 Aug 2023); www.theguardian.com/world/2023/aug/25/emperor-penguins-thousands-of-chicks-in-antarctica-likely-died-due-to-record-low-sea-ice-levels

Regaiolli, B., Sandri, C., Rose, P. E., et al., 'Investigating parental care behaviour in same-sex pairing of zoo greater flamingo (*Phoenicopterus roseus*)', *PeerJ*, 6 (2018); doi:10.7717/peerj.5227

Rézouki, C., Tafani, M., Cohas, A., et al., 'Socially mediated effects of climate change decrease survival of hibernating Alpine marmots', *J Anim Ecol*, 85 (2016); doi:10.1111/1365-2656.12507

Roenneberg, T., Aschoff, J., 'Annual rhythm of human reproduction: I. Biology, sociology, or both?', *J Biol Rhythms*, 5, 3 (1990), 195–216; doi:10.1177/074873049000500303

Smiseth, P.T., Kölliker, M., Royle, N.J., 'What is parental care?', in: Royle, N.J., Smiseth, P.T. (eds.), *The Evolution of Parental Care* (Oxford, 2012; Online Edition, Oxford Academic, 17 Dec 2013); doi:10.1093/acprof:oso/9780199692576.003.0001

Surbeck, M., Boesch, C., Crockford, C., et al., 'Males with a mother living in their group have higher paternity success in bonobos but not chimpanzees', *Curr Biol*, 29, 10 (2019), 354–355; doi:10.1016/j.cub.2019.03.040

Tokuyama, N., Toda, K., Poiret, ML. et al., 'Two wild female bonobos adopted infants from a different social group at Wamba', *Sci Rep*, 11, 4967 (2021); doi:10.1038/s41598-021-83667-2

Trumbo, S.T., 'Patterns of parental care in invertebrates', in: Royle, N.J., Smiseth, P.T. (eds.), *The Evolution of Parental Care* (Oxford, 2012; Online Edition, Oxford Academic, 17 Dec 2013); doi:10.1093/acprof:oso/9780199692576.003.0005

Twiss, S.D., Thomas, C., Poland, V., et al., 'The impact of climatic variation on the opportunity for sexual selection', *Biol Lett*, 3, 1 (2007), 312–315; doi:10.1098/rsbl.2006.0559

Valentine, K., Cross, R., Cox, R., et al., 'Caddisfly larvae are a driver of plastic litter breakdown and microplastic formation in freshwater environments', *Environ Toxicol Chem*, 41, 12 (2022), 3058–3069; doi:10.1002/etc.5496

Vincze, O., Kosztolányi, A., Barta, Z., et al., 'Parental cooperation in a changing climate: fluctuating environments predict shifts in care division', *Global Ecol Biogeogr*, 26 (2017), 347–358; doi:10.1111/geb.12540

Visser, M.E., Noordwijk, A.J. van, Tinbergen, J.M., et al., 'Warmer springs lead to mistimed reproduction in great tits (Parus major)', *Proc R Soc Lond B*, 265, 1408 (1998); doi:10.1098/rspb.1998.0514

Weaver, J., 'Parental care linked to homosexuality', *Nature* (9 Jul 2010); doi:10.1038/news.2010.344

Welbergen, J., Booth, C., Martin, J., 'Killer climate: tens of thousands of flying foxes dead in a day', *The Conversation* (24 Feb 2014); www.theconversation.com/killer-climate-tens-of-thousands-of-flying-foxes-dead-in-a-day-23227

Wilkinson, G.S., Carter, G.G., Bohn, K.M., et al., 'Non-kin cooperation in bats', *Philos Trans R Soc Lond B Biol Sci*, 371 (2016); doi:10.1098/rstb.2015.0095

Winkel, W., Hudde, H., 'Long-term trends in reproductive traits of tits (*Parus major, P. caeruleus*) and pied flycatchers *Ficedula hypoleuca*', *J Avian Biol*, 28, 2 (1997), 187–190; doi:10.2307/3677313

Wong, W., 'Gay male penguins steal lesbian couple's eggs at Dutch zoo', *NBC News* (23 Oct 2020); www.nbcnews.com/feature/nbc-out/gay-male-penguins-steal-lesbian-couple-s-eggs-dutch-zoo-n1244575

Yikra, B., 'Cichlid fish that brood in their mother's mouth sometimes get eaten', *Phys Org* (9 Nov 2022); www.phys.org/news/2022-11-cichlid-fish-brood-mother-mouth.html

Young, L.C., VanderWerf, E.A., 'Adaptive value of same-sex pairing in Laysan albatross', *Proc R Soc B* (2014); doi:10.1098/rspb.2013.2473

Young, L.C., Zaun, B.J., Vanderwerf, E.A., 'Successful same-sex pairing in Laysan albatross', *Biol Lett*, 4 (2008), 4323–4325; doi:10.1098/rsbl.2008.0191